T0324811

machine learning approaches
to bioinformatics

SCIENCE, ENGINEERING, AND BIOLOGY INFORMATICS

Series Editor: Jason T. L. Wang
(New Jersey Institute of Technology, USA)

Published:

Vol. 1: Advanced Analysis of Gene Expression Microarray Data
(Aidong Zhang)

Vol. 2: Life Science Data Mining
(Stephen T. C. Wong & Chung-Sheng Li)

Vol. 3: Analysis of Biological Data: A Soft Computing Approach
(Sanghamitra Bandyopadhyay, Ujjwal Maulik & Jason T. L. Wang)

Vol. 4: Machine Learning Approaches to Bioinformatics
(Zheng Rong Yang)

Forthcoming:

Vol. 5: Biodata Mining and Visualization: Novel Approaches
(Ilkka Havukkala)

machine learning approaches to bioinformatics

zheng rong yang

University of Exeter, UK

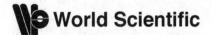

World Scientific

NEW JERSEY · LONDON · SINGAPORE · BEIJING · SHANGHAI · HONG KONG · TAIPEI · CHENNAI

Published by

World Scientific Publishing Co. Pte. Ltd.

5 Toh Tuck Link, Singapore 596224

USA office: 27 Warren Street, Suite 401-402, Hackensack, NJ 07601

UK office: 57 Shelton Street, Covent Garden, London WC2H 9HE

British Library Cataloguing-in-Publication Data
A catalogue record for this book is available from the British Library.

Science, Engineering, and Biology Informatics — Vol. 4
MACHINE LEARNING APPROACHES TO BIOINFORMATICS

Copyright © 2010 by World Scientific Publishing Co. Pte. Ltd.

All rights reserved. This book, or parts thereof, may not be reproduced in any form or by any means, electronic or mechanical, including photocopying, recording or any information storage and retrieval system now known or to be invented, without written permission from the Publisher.

For photocopying of material in this volume, please pay a copying fee through the Copyright Clearance Center, Inc., 222 Rosewood Drive, Danvers, MA 01923, USA. In this case permission to photocopy is not required from the publisher.

ISBN-13 978-981-4287-30-2
ISBN-10 981-4287-30-X

Printed in Singapore.

Preface

Bioinformatics has been one of the most important multidisciplinary subjects in the last century. Initially, the major task of bioinformatics research was to handle large genomic data for knowledge extraction and for making predictions. More recently, the practices of bioinformatics have extended from genomics to proteomics, metabolomics, and most importantly systems biology. In addition to most traditional bioinformatics exercises which focus on large database management and sequence homology alignment for molecular structure prediction and function annotation, modelling biological data using statistical/ machine learning has been an important trend. This part of the exercise has gained great attention because it can help carry out efficient, effective, and accurate knowledge extraction and prediction model construction. However, the application of machine learning approaches in bioinformatics researches and practices has a series of challenges compared with other applications. The challenges include data size, data quality, and the imbalance between different data resources. These challenges are particularly obvious in systems biology research. For instance, genomics data size has a scale of around 25K, but proteomics data size can reach up to a scale of millions. Currently, it is hard to use modern computers to handle such large scale data in one machine learning model. Furthermore, due to experimental variation, tissue corruption, and equipment resolution, most metabolite data suffer a problem of data quality. This casts a challenge in machine learning model construction in terms of data noise and missing data. In using next generation sequencing equipment such as Illumina, we are faced with tega-byte of fragments of sequences. The challenge is how to assembly these fragments accurately without any reference sequences. An urgent requirement in systems biology proposes to use different sources of data for analysing systems behaviour. This then casts a challenge about how to efficiently incorporate these data with different resolutions, with different data format, with different data quality, and with different data dimensionalities in one machine learning model. This book therefore tries to discuss some of these challenges.

This book is written based on my teaching and research notes in bioinformatics in the past ten years. I thank Prof Jason Wang and the publisher for inviting me to write this book. The book is written mainly for postgraduates and researchers at the start of their bioinformatics research and practice. The pre-requisite to using this book is some basic linear algebra and statistics knowledge. The book can be used for both advanced undergraduate and postgraduate teaching reference. Readers are encouraged to be familiar with basic R programming before using this book as most case studies presented in the book are implemented in R.

The book is composed of three parts. The first part covers several unsupervised learning approaches which can be used in bioinformatics. For instance, multidimensional scaling is commonly used in bioinformatics for biological data visualisation. Various cluster analysis approaches as well as self-organising map have been used for biological pattern recognition. After data partitioning, molecules can then be clustered leading to prototype pattern discovery and new hypothesis generation.

The second part mainly discusses supervised learning approaches. In many bioinformatics projects, a typical question is how to accurately predict unknowns based on experimental data. For instance, how can we identify the most important genes for most efficient and accurate disease diagnosis? Additionally, given a huge number of molecular sequence data in which most functions are still unknown, how can we make prediction models based on limited information of known functions in sequence data? This part therefore introduces several commonly used supervised learning algorithms as well as their applications to bioinformatics.

The third part of this book introduces the concepts relevant to computational systems biology which is now the most important research targets in bioinformatics. Computational systems biology research mainly focuses on large biological systems aiming to reveal the complex interplay between molecules and molecular entities. Gene network, systems dynamics and pathway recognition have been of much interest in recent years. The third part then demonstrates how machine learning algorithms can be used for these issues.

As mentioned above, this book is based on the revision of my teaching and research notes. It is therefore important to name several research collaborators. My key research collaborators include T Charlie Hodgman, Andrew Dalby, Murray Grant, Richard Titball, Nick Smirnoff, and Tom Richards. The students who have contributed to the improvement of my teaching of bioinformatics in University of Exeter are Rebecca Hamer, Jon Dry, Emily Berry, Dave Trudgun, Hanieh Yaghootkar and Susie Clark. I am very grateful to Susie Clark for proof-reading the book.

Finally, I would like to thank my parents, wife and daughter for their great support. During the writing of this book, I regret not being able to spend more time with them. I hope the publication of this book will make up for the sacrifice.

Zheng Rong Yang
29 November 2009
Exeter, England, UK

Contents

Chapter 1

Introduction

Bioinformatics has been in action for at least three decades. However, there is still a general confusion as to the function of bioinformatics. Some biologists are still treating bioinformatics as tools. Some informatists[1] regard bioinformatics as a career of developing novel algorithms and systems. Because of this, there is a slight difference in definitions. In the literature, one fundamental concept is also missing: that information is a natural, inherent, and dynamic component in all biological systems.

We first examine how bioinformatics is defined in various textbooks. In Attwood and Parry-Smith's book [1] bioinformatics is defined as "the application of computers in biology sciences and especially analysis of biological sequence data". In Baxevanis and Ouellette's book [2] bioinformatics is "a field integrating molecular biology and computational methods". In Higgs and Attwood's book [3] bioinformatics is defined as "the use of computational methods to study biological data". In Baldi and Brunak's book [4] bioinformatics is "the development and application of computer methods for analysis, interpretation, and prediction, as well as the design of experiments". In Mount's book [5] bioinformatics is defined as "the application of computational methods to DNA and protein science". In Augen's book [6] bioinformatics has been extended to include "*in silico* molecular modelling, protein structure prediction, and biological systems

---

[1] I use informatists to refer to a group of scientists who have the skills to apply the fundamental concepts in computer sciences, applied statistics, applied mathematics, and engineering to generate models.

1

modelling". Finally, one of the important concepts in biological research (relationship) has been used in Eidhammer, Jonassen and Taylor's definition [7], that bioinformatics is "the study of biological information and biological systems – such as the relationship between the sequence, structure and function of genes and proteins".

We then examine the definitions according to dictionaries and organisations. The Oxford English Dictionary defines bioinformatics as "the science of collecting and analysing complex biological data such as genetic codes". According to NIH, bioinformatics is defined as "research, development, or application of computational tools and approaches for expanding the use of biological, medical, behavioral or health data, including those to acquire, store, organize, archive, analyze, or visualize such data". The National Center for Biotechnology Information, defines bioinformatics as "the field of science in which biology, computer science, and information technology merge into a single discipline." NCBI also notes three important sub-disciplines within bioinformatics. The first is the development of new algorithms and statistics for accessing relationships among molecules of large data sets. The second is to analyse and interpret various data types. The outcome of these two is the integration of molecules into systems. This is also the basis of systems biology. The third is to develop and implement tools for efficient access and management of different types of information. This covers various web services and tools for public use. Both NIH and NCBI definitions cover a wide range of activities in bioinformatics.

I have no intention of giving a unique definition of bioinformatics. First, this is unfair for a huge diversity of research interests and points of views in bioinformatics. Second, the field of bioinformatics is still progressing rapidly. Many new methodologies are being developed. This book would like to treat *current* bioinformatics as a multi-discipline, inter-discipline, and cross-discipline science for understanding biological systems, exploring underlying mechanisms of biological complexes, verifying biological hypotheses and providing evidence through *in silico* simulation for further theoretical development. The requirements for bioinformatists should not be passively taking part in biological research projects. Instead, they should possess basic multi-disciplinary knowledge to undertake biological research activities independently leading to scientific findings. It is expected that wet laboratory and dry laboratory (*in silico* simulation) will become inseparable in the future for biosciences research.

1.1 Brief history of bioinformatics

Bioinformatics has generally gone through four major stages. In the *first* stage some small-sized databases and fundamental concepts for analysing sequences were established. The theoretical work of some great bioinformaticians laid the foundations. In the *second* stage, sequence analysis algorithms and programs as well as some moderate-sized databases were established. Along with the development of the internet, web services appeared. In the *third* stage, bioinformatics was not solely a market for sequence analysis. The analysis of other molecular data started, such as gene expression data and metabolite data in many medical applications. If we treat the second stage as the stage for natural finding (DNA discovery, protein structure/function annotation and many other hypothesis-based projects), this stage is more application-driven. Many bioinformatics projects have wide support from industry and medical services. The *fourth* stage is for systems-level examination of biological systems. This is a natural development from the third stage where it is difficult to gain a complete picture by analysing individual cases. Integrating molecules from the same data type or different data types has been an urgent task for un-biased understanding of cellular activities.

When looking at the history of bioinformatics, two important pioneering works must be remembered. The first is Pauling and Zuckerkandl's molecular evolution theory developed in the early 1960's [8, 9]. The work illustrated that amino acid sequences of proteins can be used to study evolutionary relationships among organisms. They showed that two proteins with homologous amino acid sequences have similar functions. The work therefore initiated a new field known as "molecular evolutionary". The theory provides theoretical basis for inferring protein functions based on sequence homology. The technique is call homology alignment [10-15].

The second important work is the computerised protein and DNA sequence databases of Margaret Oakley Dayhoff in the 1970's based on her knowledge of chemistry, mathematics, biology and computer science. From this, she derived evolutionary histories using sequence homology with Pauling and Zuckerkandl's theory. She developed phylogeny for the first time with Richard Eck [16, 17]. The first probability model of protein evolution, referred to as point mutation process, was also her contribution [18]. Her quantitative measure of protein evolution, known as the mutation matrix [15], has been widely used in today's bioinformatics tools.

Based on the successes of Pauling and Dayhoff, rapid progress in bioinformatics started in the 1970's because of the rapid technology development in computers. The progress mainly focused on DNA and protein sequence analysis. Because of the time complexity, the main focus was on improving algorithm speed especially for sequence homology alignment. The comparison of genes within a species or between different species can be used to indicate structural and functional similarity. In 1970, the first sequence homology alignment algorithm was developed and is referred to as the Needleman-Wunsch algorithm [19]. The algorithm aligns two sequences globally using a dynamic programming approach. In this algorithm the comparison between two sequences is based on a binary scoring function. The score is increased by one when the current aligned residues from two sequences match, otherwise zero. In addition, linear gap penalty is used. In the algorithm insertion and deletion is considered. Therefore two sequences with different lengths will be aligned to the same length with inserted gaps. As seen above, all the matching residues have the same score as one and all the mismatching residues have the same score as zero. The first computer program for DNA sequencing was developed in 1977 [20]. The program can be used for effectively assembling sequence data. In 1981, an important concept called sequence motif for sequence analysis was generated [21]. In the same year the Smith-Waterman algorithm was developed [13]. The algorithm also aligns two sequences using a dynamic programming approach to guarantee finding the optimal local alignment with respect to the substitution matrix and the gap penalty function used. The algorithm is a local alignment algorithm, which is due to the difficulty of obtaining correct alignments in regions of low similarity between distantly related biological sequences. However, the Smith-Waterman algorithm is a slow algorithm requiring a large memory. Because of this, it has been replaced by much more efficient algorithms for instance the FASTP algorithm published in 1983 [14], the FASTP/FASTN algorithm published in 1985 [22] and the BLAST algorithm implemented in 1990 [23].

Contributing to the third generation of bioinformatics are vast activities in analysing gene expression data. A gene is the basic unit of heredity in all living organisms; it is a segment of DNA sequence, a unit coding genetic information which is inheritable [24-26]. In other words, DNA is an organisation of information [27]. Genes are transcribed to RNAs which in turn are translated to proteins. This is controlled by a gene regulation process [24-26]. Gene expression is a process whereby a relevant gene is transcribed and translated to RNAs and proteins respectively according to a regulatory signal. These RNAs and proteins are functional in certain pathways or networks. Gene expression can be measured quantitatively using biotechnology. The measurements can be at the RNA level or protein level depending on techniques used. It is understood in molecular biology that a specific pattern of gene expression in a number of biologically related samples represents the activity of a specific signalling pathway or network. The bioinformatics study of gene expression data was triggered by the generation of DNA microarray data in the 1980's. A DNA microarray is a technology developed particularly for medicine. Each microarray is an array of thousands of DNA oligonucleotides from biologically relevant samples. The samples can be related to a specific disease diagnosis. One group of samples can be disease-free and the other can be disease-related. By analysing the pattern of expression of these DNA oligonucleotides, it is possible to investigate the genetic reason of disease development. Microarray technology evolved from Southern blotting [28] and the first use of DNA microarray expression profiling was in 1987 for identifying genes whose expression is modulated by interferon [29]. The earliest report in analysing microarray expression data of the budding yeast

Saccharomyces cerevisiae using cluster analysis approach was in 1998 by Eisen *et. al.*, [30]. Recent studies in clustering microarray expression data include those looking at renal cell carcinoma [31], inflammatory immune signalling in chronic fatigue syndrome [32], inflammation status in hepatitis C virus-related hepatocellular carcinoma [33], etc. Classification models have also been built for predictive/diagnostic purposes, such as the diagnosis of breast cancer [34], [35], colorectal cancer [36], lung cancer [37], brain cancer [38], ovarian cancer [39], etc.

In the fourth generation of bioinformatics, many researchers turn their eyes to systems biology, which is an inter-discipline and cross-discipline subject in studying biological systems. The major objective of systems biology is to discover new emergent properties of processes at the cellular level and organism level in biological systems in a systematic view. Following this, a number of systems biology institutes have been established and some doctorial training centres have also been created. Although the huge scale of systems biology studies started only a decade ago, the earliest work using the systems biology approach to study biological processes was published in the 1950's [40]. A foundation study of systems biology was completed in the 1960's with the publication of Mesarovic's book [41]. The first systems biology institute was established in 1999 [42] in the Department of Molecular Biotechnology at the University of Washington, aiming to model complex biological systems quantitatively and foster interdisciplinary interactions in the life sciences.

1.2 Database application in bioinformatics

The introduction of database technology into bioinformatics in the early days was brought about by the development of many gene/protein sequencing projects which needed an efficient way for data handling. In the 1930's, electrophoresis was developed for separating proteins in solution using moving boundary or zone electrophoresis [43]. The structure of the alpha-helix and beta-sheet was proposed in the early 1950's [44, 45] and the double helix model for DNA based on x-ray experiment was proposed in 1953 [46]. The first sequenced protein (bovine insulin) was analysed in 1955 [47]. Herbert Boyer and Stanely Cohen invented DNA cloning or recombinant DNA technology in 1973 [48]. The technology made it possible to manipulate DNA in different species. For instance, some parts in DNA can be removed or replaced and some altered segments can be inserted into DNA. Specific proteins can be produced using gene splicing. In order to analyse the presence of a DNA sequence in a DNA sample the Southern blot was developed in 1975 [28]. The first sequenced DNA was seen in 1977 [49, 50]. In 1980, a multi-dimensional NMR method was developed for protein structure determination [51]. In 1996, the first DNA chip was generated by Affymetrix (NASDAQ: AFFX). The first gene Chip product was an HIV genotyping GeneChip. The human genome with 3000 Mbp was produced in 2004 [52]. Based on this simple description of molecular data generation history it can be seen that, on the one hand, technologies are fast developing and, on the other hand, data sizes are dramatically increased, making a huge challenge for data handling, management, mining, i.e. bioinformatics.

In order to fulfil the needs in acquiring data for research, various databases continue to be established thereafter. In 1986, the largest curate protein databank SWISS-PROT was created by the Department of Medical Biochemistry of the University of Geneva and the European Molecular Biology Laboratory (EMBL). In 1988 The National Center for Biotechnology Information (NCBI) was established at the National Cancer Institute. Many successful projects of building data warehouses have well used and well developed database technology in computer sciences for a huge amount of molecular data. Efficiently storing sequence data is one important topic. A number of nucleotide sequence databases and protein sequence databases have therefore been implemented. The well-known nucleotide sequence databases include GenBank [53, 54] referred to as the NIH genetic sequence database, EMBL Nucleotide Sequence Database [55] referred to as the European equivalent to the U.S.'s GenBank, DDBJ (DNA Data Bank of Japan), Human Genome Sequencing Centre at Baylor College of Medicine, IMGT (the International ImMunoGeneTics Database) [56]. The widely used protein sequence databanks are UniProt (United Protein Databases) and Swiss-Prot. The UniProt is a centralised database cooperating with

EBI (European Bioinformatics Institute), PIR (Protein Information Resource), GUMC (Georgetown University Medical Centre), NBRF (National Biomedical Research Foundation), and SIB (Swiss Institute of Bioinformatics). The Swiss-Prot is the major European protein sequence database. In Swiss-Prot, various properties of proteins are stored such as the description of the function of a protein, protein domains structure data, and protein posttranslational modifications data.

1.3 Web tools and services for sequence homology alignment

Since DNA and protein sequencing technologies have been successfully developed, many DNA and protein sequences have been well organised and stored in various databases as mentioned above. One of the urgent tasks is to have tools which can compare two sequences to indicate how similar they are. Based on well-developed homology alignment algorithms, web tools have been developed and are open to the public. For instance, some BLAST tools are implemented in the National Center for Biotechnology Information (NCBI): nucleotide blast, used for searching a nucleotide database for a nucleotide query based on the BLASTn algorithm, protein blast, used for searching the protein database for a protein query based on BLASTp, Position-Specific Iterated – BLAST or psi – BLAST [57, 58], Pattern Hit Iterated – BLAST or phi – BLAST [59], BLASTx, tBLASTn, and tBLASTx. Most of these have been implemented as web tools. All of them deal with predictions indirectly. For instance, a query sequence may have been aligned with a number of database sequences. These database sequences have known structures and functions. If the query sequence has a high returned similarity with these database sequences, the conserved segment corresponding to protein structures or functions in these database sequences can be used for the prediction of the query sequence. A web tool will enable the user to enter a query on the internet while a server of a web tool will conduct all the necessary computing. The computing result will be returned to the user either on the web site or by an email.

The FAST/BLAST series tools are used for aligning a query sequence against many database sequences to find the most similar ones. The algorithms implemented in all tools consider insertions and deletions. There are also two other classes of web tools implemented in bioinformatics studies, one being prediction using whole protein sequences and another being prediction using sub-sequences or peptides. These two classes of web tools are used for direct protein function prediction. For instance, the tools developed for the prediction of protein localisation [60-62], gene structure prediction [63] and function annotation [64] use whole protein sequences as input to predict protein structures and functions directly.

1.3.1 *Web tools and services for protein functional site identification*

Protein functional site identification using peptides includes the prediction of protein cleavage sites, protein posttranslational modification sites, binding sites, and turn types. For instance, bioinformatics algorithms and (web) tools have been used to predict proteasomal cleavage sites [65], promiscuous MHC Class-I binding sites [66], RNA binding sites [67, 68], lipoprotein signal sites [69], transcription binding sites [70], active sites [71], ligand binding site [72], miRNA target site [73], protein-protein interaction sites [68], convertase sites [74], SH3 domain interaction sites [75], and signal peptides [76].

In predicting posttranslational modification sites, there are also many web tools being developed, for instance, glycosylation site prediction [77, 78], phosphorylation site prediction [74, 79-83], acetylation site prediction [83], methylation site prediction [84], sumoylation site prediction [85], palmitoylation site prediction [86] and GPI-modification site prediction [87]. Web tools have also been implemented for protein turn prediction [88-90]. Another class of web tools for protein structure prediction uses variable peptide length for prediction. This class of web tools include protein disorder prediction and secondary structure prediction. For predicting secondary structures in proteins, the implemented web tools are PreSSAPro [91], E-SSpred [92] and MUPRED [93], PROTEUS [94], GOR V [95], Porter [96] and logic alignment approach [97]. MeDor [98], DPROT [99], iPDA [100], PrDOC [101], FoldUnfold [102], Spritz [103], IUPred [104], RONN

[105], DisEmbl [106], TOP-IDP-scale [107], GlobPlot [108] and PONDA [109] are the web tools for disordered protein prediction.

1.3.2 *Web tools and services for other biological data*

Web tools have also been implemented for other biological data analysis, for instance for RNA data analysis [110], RNA deleterious mutation analysis [111], microarray data interpretation [112], transcriptional regulatory network construction [113] and for gene selection and classification [114], [115]. Web services also cover metabolite data analysis, such as correlating ligand metabolites with pathways [116] and integrating transcripts and metabolites [117]. All these efforts aim to help biologists to enhance their biological experiments and speed up scientific findings.

1.4 Pattern analysis

The third important practice in bioinformatics is pattern analysis. It covers a wide range of topics, methodologies and algorithms. This book will mainly focus on this practice providing a broad introduction and analysis. Compared with the other two subjects mentioned above, pattern analysis deals with many fundamental issues in bioinformatics. If a web tool is more or less computing technique-based, pattern analysis needs some fundamental support from statistics and mathematics. From this, models or web tools can be constructed.

Pattern analysis focuses on the exploration of the underlying mechanism of biological data. It aims to find the rules which govern data distribution. Only by knowing these rules, can proper models be constructed. For instance, in any prediction system, the most important part is a prediction model. Without fully understanding how data are distributed, no accurate or efficient model can be constructed for prediction. In order to build a proper predictor, a rigorous modelling process based on statistical modelling principles must be followed.

Pattern analysis mainly involves two learning mechanisms, i.e. unsupervised learning and supervised learning. The former is for knowledge discovery, rule extraction and data visualisation, while the latter is for predictive model construction. There are also many different algorithms for each learning mechanism, some being simple leading to coarse but easy-to-interpret models, some being complicated leading to some accurate but difficult-to-interpret models.

In recent years, systems biology and computational systems biology have been paid increasing attention because of their importance in understanding biological systems. Conventionally, biological studies often decompose a system into some very basic and small systems. The study of these decomposed systems may miss important information of complex interplay in cells or organisms. Two trends have emerged in systems biology study. They are top-down compositional analysis, aiming at predicting system dynamics, and bottom-up integrating analysis, aiming at putting molecules into the right classes, pathways, or networks.

1.5 The contribution of information technology

The development, progress, and advances of bioinformatics could never have taken place without the support of IT successes. In 1946, came the announcement of the Turing-complete, a digital computer [118]. It is referred to as Electronic Numerical Integrator And Computer (ENIAC). The main purpose of ENIAC was to calculate artillery firing tables for the U.S. Army's Ballistic Research Laboratory although it can be used to solve various computer programming problems. The advantage of ENIAC is its speed: one thousand times faster than an electro-mechanical machine. Meanwhile, its power in dealing with mathematics for general-purpose programming promoted the spread of using computers in various applications.

In 1958, another revolution occurred in electronics which is closely related to the computer industry. The event was the development of the integrated circuit (IC) which made the manufacture of electronic equipments much faster and cheaper. Later, IC quickly progressed to very large scale IC (VLSI) leading to almost all electronic equipments including computers in use today being packed into a very small space.

Particularly, VSLI has greatly improved the efficiency of the core parts of a computer (CPU – central processing unit) in two ways. First, the size of CPU can be much smaller. Second, the memory is dramatically increased.

Because of the huge progress in electronics and computers (nowadays referred to as hardware in contrast to programming codes as software), using computers to store sequence data has become a convention. However, the following events have also made bioinformatics research feasible.

In 1969, Unix systems appeared in the Bell laboratory, which provided a powerful platform for large scale computing. The next important event was the emergence of the internet. The first internet (1st generation) was called Advanced Research Projects Agency Network (ARPAnet) established by the United States Department of Defence. ARPAnet was first established on November 21, 1969 linking the IMP at UCLA and the IMP at SRI. The 2nd generation was connecting desk PCs through telephone lines. The 3rd generation was using wireless connections to laptop computers. The 4th generation (the current one) is using mobile phone internet through cellular networks [119].

Two important network applications are email and file transfer. Email was invented in 1971. File transfer protocol (ftp) was invented in 1973. These two applications have become the most important composition parts of modern bioinformatics services. Almost all the web services and tools mentioned above include these two applications.

The other important developments in computer sciences include personal PC, window systems, Linux, Netscape, Perl programming language, Java and Java Script Programming languages; all have played important roles in promoting fast bioinformatics progress and development.

1.6 Chapters

This book is composed of 18 chapters. Except for chapters 1 and 20, the rest are divided into three parts. Chapters 2, 3, 4, 5, and 6 constitute part 1 and mainly discuss the issue of unsupervised learning. Chapter 2 introduces general concepts of unsupervised learning. Chapters 3, 4, 5, and 6 separately discuss most commonly used approaches, namely probability density estimation, principal component analysis, cluster analysis, multi-dimensional scaling, and self-organising map. Although they have some overlapped functions, each uses a distinct statistical assumption about data. All these four approaches can be used for different aspects of knowledge discovery. Chapters 7, 8, 9, 10, 11, 12, and 13 constitute part 2 and are used to cover supervised learning algorithms including linear/quadratic discriminant analysis, K-nearest neighbours, decision trees, neural networks, vector machines, and hidden Markov models. Specifically, chapter 13 focuses on an important issue in handling biological data, i.e. feature or variable selection. Chapters 14 and 15 constitute additional components for part 2 and will focus on peptide classification or functional site prediction problems. Chapters 16 and 17 constitute part 3 and will discuss computational systems biology studies including causal networks and S-systems. Chapter 18 discusses the future research directions.

Chapter 2 will focus on the general concepts of knowledge discovery approaches in bioinformatics. The chapter will discuss the principle of unsupervised learning approach and briefly introduce various unsupervised learning algorithms. The chapter will also introduce some applications of using unsupervised learning approaches to explore knowledge from large-scale biological data. Chapter 3 will introduce a useful approach in statistical learning, i.e. probability density estimation for most data analysis projects. This approach is commonly used as primary data analysis aiding proper selection of modelling algorithms. Various algorithms and procedures will also be discussed. Chapter 4 will introduce principal component analysis (PCA) and the Sammon mapping algorithm for biological data dimension reduction. PCA can lead to two outcomes, data reduction and data visualisation. In bioinformatics, PCA is commonly used to visualise data using the first and second principal components. Chapter 5 will discuss how to partition biological data through the use of various clustering algorithms. Data partitioning is commonly used in bioinformatics to visualise how data are clustered. From this, typical biological functions can be extracted. Chapter 6 will introduce the self-organising map as a neural learning algorithm which is capable of visualising, clustering data and reducing dimensionality of data.

Chapter 7 will briefly discuss the use of supervised learning approaches in bioinformatics. Some linear algorithms will be discussed first followed by nonlinear algorithms. Chapter 8 discusses linear/quadratic discriminant analysis and K-nearest neighbour algorithm as simple learning algorithms. Chapter 9 will discuss decision trees and the random forest algorithm as well as their applications to bioinformatics for exploring human-like decision-making systems. Chapter 10 will discuss neural networks which are one of the powerful nonlinear algorithms. Because neural networks have been widely used in bioinformatics applications, various cases will be discussed. Chapter 11 will discuss recent development in nonlinear classification approaches including basis function neural networks, support vector machine and relevance vector machine. Because they have the advantage of better generalisation capability and interpretation using support/relevance vectors, their applications to bioinformatics projects have gained an increasing interest. Chapter 12 will discuss hidden Markov models which have been intensively used in sequence analysis. Chapter 13 will introduce various approaches of feature selection which are critical in analysing biological data such as gene expression and metabolite data for extracting the most informative biomarkers.

Chapter 14 will discuss the coding problem which is important to the analysis of sequence data, where residues are commonly non-numerical attributes. Several coding mechanisms will be discussed and compared. Chapter 15 will focus on one specific subject in bioinformatics, i.e. peptide classification where the main topics including data selection, organisation, target definition, and modelling procedures.

Chapter 16 will discuss how to use causal network principle and Bayesian network for constructing gene networks. Chapter 17 will discuss the developments in computational systems biology. The focus will be mainly on metabolite data analysis. Chapter 18 will outline the future research directions in bioinformatics.

Chapter 2

Introduction to Unsupervised Learning

In many real-world applications, available data may have little domain knowledge (signature) associated, for instance, the data structure and the inference rule of a data set may be missing or yet to be discovered. Such a data set is categorised as incomplete data for which inference on novel data becomes difficult. In order to make data, particularly experimental data, useful for inference it is necessary to explore signatures for a data set, which should fit well the inference purposes. For instance, a mass spectrometry experiment on a set of plant samples can generate many thousands metabolites. Each metabolite is represented by a mass and a number of abundance values for replicates. Based on masses we can infer a number of the chemical formulas of candidate compounds from different pathways. In theory, one mass corresponds to one compound. However one metabolite may be mapped to multiple compounds. A selection process is commonly conducted manually in a laboratory to identify the true compound of the metabolite. A manual verification is prone to error and is also cost demanding. An automatic process can therefore be helpful to cover these two issues. It is understood that molecules in the same pathway should have similar responses to the treatment induced in an experiment. It is possible to study the clusters of metabolites according to abundance values. If some metabolites in a cluster have definite compound/pathway annotations, a metabolite with multiple mappings in the same cluster can then be identified. Another example in studying metabolite data is the identification of a proportion of important metabolites which can discriminate between experimental and control groups of samples. To fulfil this purpose, we can model the

15

density functions of the differential intensities of metabolites from the experimental and control groups. By setting a threshold, say 1%, we can identify a subset of the most important metabolites for discrimination.

The process of exploring knowledge from data in this study is referred to as a learning process in machine learning. There are mainly two learning processes used in analysing biological data, i.e. unsupervised and supervised learning. Two learning processes adopt two different learning mechanisms. A supervised learning process seeks a mapping function from one data space to another data space. For instance, if we have enough biological knowledge, we may map chemical formulas of compounds to pathways directly, where chemical formulas and pathways are certainly in two different data spaces. The association between these two spaces is the goal to acquire in a supervised learning process. With unsupervised learning process, it is assumed that one data space is missing. For instance, a set of gene expression profiles for a specific disease may be known, but the number of inherent causative agents of that disease may be unknown. In other words, the causative agents are not observable or not easily observed. For mapping gene expression profiles to this missing space, an unsupervised learning process can be taken. The learning process is to re-organise the available data space to explore the missing space, or to map the available data to the missing space.

Supervised learning process will be detailed in Chapters 7 to 13. From this Chapter to Chapter 6, the focus is on unsupervised learning approaches. Most machine learning approaches or algorithms including unsupervised and supervised learning algorithms involve a parameter space optimisation problem. It is assumed that knowledge in data or data signatures can be quantitatively expressed by parameters. In other words, parameters are the quantitative characterisation of knowledge by which data are generated. For the gene expression data mentioned above, we may say that it is the body of the causative agents which are not observed that generate the gene expression profile data. If the causative agents vary from disease to disease, the gene expression profile will vary as a consequence. It must be noted that knowledge is an abstract object by which exhausting data are infinite. For instance, if time and cost are allowed, we may collect infinite gene expression profile data for one disease. Limited by time and cost, the data obtained for acquiring knowledge can only be one random sample of the true knowledge. For instance, given a mean and a standard deviation of the Gaussian distribution, many data sets can be randomly generated. Figure 2.1 shows an example of this, where the mean and the standard deviation of the Gaussian distribution are zero and one, respectively. Sampling the distribution twice with 100 data points generates two random data sets from which two histograms are drawn displaying two different data distributions. One of these two data sets may be used for the inference of mean and standard deviation of the Gaussian distribution.

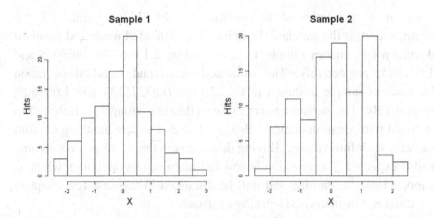

Fig. 2.1. Two samples are generated by randomly sampling the Gaussian distribution characterised by mean: zero, and standard deviation: one, with 100 data points. Two histograms are generated using two data sets. In the histogram, the range of X (the independent variable) is divided evenly into a number of intervals referred to as bins. The data falling into each bin are counted as "Hits" placed as the vertical axes in the Figure.

In unsupervised learning, such a learning process normally deals with one data set as mentioned above. In the above example, the inference of the Gaussian characterisation parameters (mean and standard deviation) will not have any other data to support. In other words, the source (a Gaussian function) which generates the data is missing. What an unsupervised learning process does is to find this information hidden in data through learning. For instance, the estimation of the mean and standard deviation of the Gaussian distribution for data samples in Fig. 2.1 can be done by

$$\mu = \frac{1}{n}\sum_{i=1}^{n} x_i \tag{2.1}$$

and

$$\sigma = \sqrt{\frac{1}{n}\sum_{i=1}^{n}(x_i - \mu)^2} \tag{2.2}$$

where n is the number of data points, x_i is the ith data point, μ is the mean, and σ is the standard deviation. The estimated mean and standard deviation for random sample 1 shown in Fig. 2.1 are -0.01468948 and 1.051885, respectively. The estimated mean and standard deviation for random sample 2 shown in Fig. 2.1 are 0.07375137 and 1.008609, respectively. The parameters are close to the true parameters (referred to as truth) with small deviations. Bear in bind that we are handling random samples with limited size. Having deviations is then a common outcome and is expected. The question is how to minimise this deviation, which is another issue in learning and will be discussed in the next few chapters for different unsupervised learning approaches.

In the above example, the Gaussian distribution is regarded as a data structure while the parameters are regarded as the inference rule. Unless these two parameters are well-learnt, an inference process may not be accurate and correct. The correctness means that an inference must find the correct data structure while accuracy measures how close the estimated parameters are to the true parameters. The assumed data structure for two random samples in Fig. 2.1 is Gaussian. If the assumed data structure is incorrect, the explored inference rule will be useless. For instance, the data available (in solid line in Fig. 2.2) do not follow the Gaussian distribution. If we assume that the data are generated from a Gaussian distribution, the estimated inference rule will not generate any useful prediction. The dotted line in Fig. 2.2 shows a significant deviation from the solid line. This incorrect information will be useless or misleading in future inference.

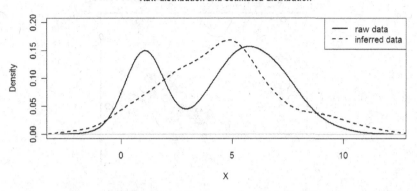

Fig. 2.2. An example that uses an incorrect data structure for inference. The solid line is the density function (details of this will be discussed in Chapter 3) of a random sample generated from a true data structure which is composed of multiple Gaussian distributions. The dotted line is the density function of an estimated inference rule of the assumed Gaussian data structure.

The above two examples are categories of density estimation in machine learning. As stated above, it is necessary to explore the true data structure as well as the true inference rule.

There are also two other important subjects of unsupervised learning in machine learning. They are data visualisation and cluster analysis. Data visualisation is a powerful tool in real data analysis where the main objective is to investigate how data are distributed. From this, further studies can follow that look at data structures and inference rules. If data are located in a low-dimensional space, such an investigation will not be very difficult. However, in most real bioinformatics applications, data are sitting in high dimensional spaces in which it is impossible to visualise how data are distributed. For instance, microarray gene expression data is a typical example. Gene expression data for studying certain diseases may be available in only a few samples but may have thousands or hundreds of thousands of genes used as variables. Direct and intuitive visualisation of the data is impossible unless a specific treatment is taken. In order to visualise high-dimensional data we then need to use various visualisation approaches. Some are simple while some are tailored for handling nonlinear and complicated data.

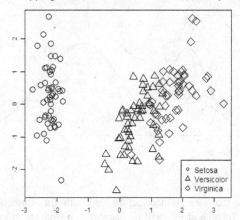

Fig. 2.3. Mapping the Iris data (in four dimensions) to a two-dimensional space. There are three species of Iris flowers; each has four descriptions, hence four-dimensional data space.

Figure 2.3 shows a visualisation map of Iris data in which three species of Iris flowers (Setosa, Versicolor, Virginica) are quantified by four variables (sepal.length, sepal.width, petal.length, and petal.width). In using these four variables, a map is generated using a visualisation algorithm, a multi-dimensional scaling algorithm which will be discussed in Chapter 4. In the map, it can be seen that the species Setosa is well separated from the other two species, which are difficult to separate.

A data set may not have one unique data structure, i.e. a data structure can be viewed as a composition of disjointed sub-data structures. Each sub-data structure is a collection of data points with similar physical background. For instance Fig. 2.4 shows a data structure with four sub-data structures. The four contours are four sub-data structures based on which four clusters of data points are generated using random sampling. In unsupervised learning, a key issue is how to find this data structure from a random sample and quantitatively describe the data structure. Therefore two important issues are whether the data structure and the inference rule which are both acquired are estimated correctly and accurately. Such a process in unsupervised learning is called cluster analysis.

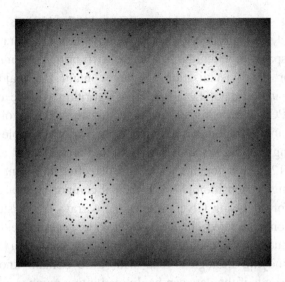

Fig. 2.4. Four sub-data structures of a data set. Four contours are for four sub-data structures and dots comprise a random data set sampled from a data structure with four sub-data structures.

These three typical unsupervised learning processes have been widely used in various Bioinformatics problems. Some are used for data pre-processing, some are used for data primary study and some are used for building predictors. For the first two purposes, unsupervised learning approaches are normally used for identifying data structures and inference rules. A further progress from this is to build predictors based on the found data structure and the estimated inference rules. For instance, if a data structure is found to have two Gaussian distributions for two classes of data, e.g. disease-related and disease-free, with an estimated optimal inference rule, e.g. the Gaussian parameters, classification or prediction can be made for novel data. Various density function estimation algorithms [120] have been used in bioinformatics, for example with the prediction of miRNA [121], the prediction of secondary structures [122], the functional annotation of proteins using gene ontology contents [123], the segmentation of cDNA microarray spots [124], the prediction of protein crystallization propensity [125], and the prediction of protein functional sites [126].

Principal component analysis (PCA) [127, 128] is a powerful unsupervised learning algorithm which converts a raw data space to an orthogonal space in which the first one or two principal components can be used for data visualisation if they contain the majority of the information in the data. Details of the algorithm will be discussed in Chapter 4. PCA has been used for detecting the aging evolution of the volatile organic compounds obtained from various samples [129], for characterising a broad range of structural and architectural alterations in cell walls [129], and for analysing and detecting different qualities of food and drug composition [130]. Principal component analysis has also been used for analysing Glycoprotein microarrays [131], studying low-dose radiation-associated changes in cytokine gene expression profiles [132], and studying multi-dimensional gene expression data [133].

When studying biological data with little domain knowledge, an important issue is if the data can be divided into a number of small, non-overlapping groups. If so, how many groups can be made for a data set? This is key to investigating whether a biological phenotype is composed of several non-overlapping genotypes or for studying the causative agents leading to a specific disease. The approach for this is called cluster analysis. Cluster analysis has been used for representing and analysing gene sequences [134], protein sequence analysis [135], deriving sequence templates so as to analyse protein tertiary structures [136], studying the genetic diversity of harpins from Xanthomonas oryzae [137], testing the importance of myogenic gene expression during myofiber hypertrophy in humans [138], analysing gene expression data of human dental pulp stem cells [139], analysing SAS on real-time PCR gene expression [140], subcategorising of tumour types through gene-expression profiling [141], studying gene expression dynamics [142], and optimising gene cluster structure using biological knowledge [143, 144].

As a neural network unsupervised learning algorithm, self-organising map (SOM) [145] has been widely used in bioinformatics because of its powerfulness in associative memory and pattern analysis. The mechanism of SOM will be discussed in Chapter 6. SOM has been used for protein structure localisation analysis [146], protein turn type prediction [147], DNA fragment taxonomic visualization and classification [148], clustering short time-course microarray data with replicates [149], peptide identification [150], secondary structure prediction [151], G-protein-coupled receptors classification without alignment [152], searching for hidden sequence signatures of eukaryotic genomes [153], and optimal HP configuration [154].

Chapter 3

Probability Density Estimation Approaches

The importance of density estimation has been discussed in the last Chapter. In this Chapter, the relevant algorithms and their applications to bioinformatics are detailed. The simplest approach, called the histogram approach, is introduced at the beginning. Further discussion then follows towards three categories. First, a parametric approach is introduced with which we assume that data follow a Gaussian distribution as a hidden or missing data structure and two parameters (a mean and a standard deviation) as the inference rule of it are estimated.

Second, two non-parametric approaches are discussed. In contrast to parametric approaches, non-parametric approaches do not try to explore the explicit form of a data structure for a data set. The inference rule is therefore algorithm-oriented. Third, a semi-parametric approach is also introduced. Unlike parametric and non-parametric approaches, semi-parametric approaches explore a flexible data structure with inference rules for future decision making. For each of these three categories, a number of applications will be discussed. The data used for learning is referred to as training data.

3.1 Histogram approach

The histogram approach is the simplest density estimation approach. In estimating a density function for a data set, each coordinate is divided into segments with fixed length. Such a segment is commonly called a bin. For instance, the coordinate of a variable $x \in [a,b]$ is divided into K non-overlapping bins each of which has the length

$$\frac{b-a}{K} \tag{3.1}$$

Each training data point is scanned to see which bin the data point falls in. The frequency of each bin is equal to the number of training data points falling into it over the total number of training data points. The frequency is treated as the inference rule. In prediction, the frequency is used as the probability to indicate how likely novel data is to fall in a certain bin. Suppose a bin is characterised by the interval $\text{bin}_i = [u_1, u_2]$ ($a \leq u_1 < u_2 \leq b$) and a frequency f_i. If a novel data point falls in this interval, i.e. $x \in [u_1, u_2]$, the frequency of the bin is then used to indicate how likely $u_1 \leq x \leq u_2$. Note that "$\in$" reads as "belonging to". From this, a histogram can be generated to visualise the data distribution. The left panel in Fig. 3.1 shows such an application. Details of data will be discussed in the next section. The histogram approach is a simple method for studying probability density functions, but it has a fatal limit in computational cost when the number of variables gets larger. For D variables, there will be K^D bins. When K=10, the number of bins will be 10^{10} when a data set has 10 variables. A further note on the histogram approach is its specificity. It is normally categorised as a non-parametric approach because it does have a property of non-parametric approaches in that no explicit data structure is used. However, it has some similarity with parametric approaches in that training data will not be kept for an inference process.

3.2 Parametric approach

Unlike the histogram approach, a parametric approach makes an assumption of data structure before estimating the probability density function of a data set. We acquire a unique data structure which is Gaussian in most applications and a relevant inference rule. After the probability density function of a data set is constructed, we discard the training data. What is left is the inference rule characterised by two parameters for a Gaussian data structure. A Gaussian data structure has mean and standard deviation parameters. If a training data set with N training data points is denoted by $\Omega = \{ \mathbf{x}_n \}_{n=1}^{N}$, $\mathbf{x}_n = (x_{n1}, x_{n2}, \cdots, x_{nd})$

(hence $\mathbf{x}_n \in \mathfrak{R}^d$) with x_{nj} as the jth element of the nth training datum in Ω. Here we follow convention to denote a d-dimensional vector by a bold-faced letter. $\{\mathbf{x}_n\}_{n=1}^N$ is read as enumerating n for $\mathbf{x}_n$ from 1 to N. With the assumption of a Gaussian data structure, a likelihood function (L) that N training data points are generated by the assumed Gaussian data structure is defined as below

$$L = \prod_{n=1}^N p(\mathbf{x}_n) \tag{3.2}$$

Here the probability density function $p(\mathbf{x}_n)$ is a Gaussian

$$p(\mathbf{x}_n) = \frac{1}{(2\pi)^{d/2}\sqrt{|\Sigma|}} \exp\left(-\frac{(\mathbf{x}_n - \mathbf{\mu})^T \Sigma^{-1}(\mathbf{x}_n - \mathbf{\mu})}{2}\right) \tag{3.3}$$

with Σ as the covariance matrix and $\mathbf{\mu}$ the mean vector. The likelihood function is determined by the parameters, i.e. the mean vector and the covariance matrix. Only when the likelihood function is maximised, the two parameters are optimised or optimally determined so that the estimated Gaussian data structure can fit the training data well. If training data are orthogonal, we can assume that the covariance matrix is diagonal

$$\Sigma = \begin{pmatrix} \sigma_1^2 & 0 & \cdots & 0 \\ 0 & \sigma_2^2 & \cdots & 0 \\ \vdots & \vdots & \vdots & \vdots \\ 0 & 0 & \cdots & \sigma_d^2 \end{pmatrix} \tag{3.4}$$

We can further assume that data are homogeneous. From this the covariance matrix becomes $\Sigma = \sigma^2 \mathbf{I}$ with $\mathbf{I}$ as an identity matrix. Applying a logarithm to the likelihood function and using the maximum likelihood approach leads to

$$\mathbf{\mu} = \frac{1}{N}\sum_{n=1}^N \mathbf{x}_n \tag{3.5}$$

and

$$\sigma^2 = \frac{1}{N}\sum_{n=1}^{N}(\mathbf{x}_n - \boldsymbol{\mu})^{\mathrm{T}}(\mathbf{x}_n - \boldsymbol{\mu})$$ (3.6)

Figure 3.1 shows an application of parametric density estimation for the hydrogen distribution in compounds. The data are from the Kegg library [155]. There are 2762 compounds. Each compound is expressed by a formula. For instance, the chemical formula of ADP is $C_{10}H_{15}N_5O_{10}P_2$ with relative mass as 427.029297. In this compound, there are 15 units of hydrogen. Among 2762 compounds, the quantity of hydrogen varies and the estimation is based on the following data conversion

$$x = \log(H + 1)$$ (3.7)

where H means the quantity of hydrogen. Using the parametric approach described above, we can estimate the probability density function for x.

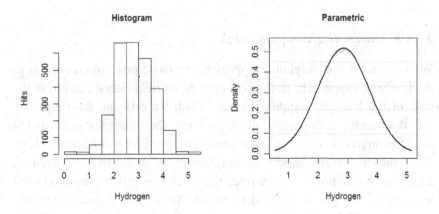

Fig. 3.1. (a) The histogram; (b) The parametric approach for estimating probability density function of hydrogen quantity in compounds.

Shown in the left panel in Fig. 3.1, a histogram demonstrates biased Gaussian distribution for comparison. It can be seen that the parametric approach delivers a symmetrical density function while the histogram shows a skewed density function. From this case, we can see that we must be careful when using the parametric approach for probability density estimation. If the difference between a real distribution and the predicted one is too large, we need to think of an alternative. The parametric approach has the advantage of being very simple and straightforward.

3.3 Non-parametric approach

A non-parametric approach in machine learning means that a model is built without a clearly defined data structure. A model will need to use all the available data points for an inference. The prediction is based on a specific inference rule replying on the relation between a novel datum and whole training data. Two commonly used non-parametric approaches for density function estimation are discussed here. They are the nearest neighbour approach and the kernel approach. The latter has been embedded in the R project.

3.3.1 *K-nearest neighbour approach*

With the K-nearest neighbour approach, the basic principle is similar to the histogram approach. Both use a predefined field (named as bin in the case of the histogram approach) into which we estimate the frequency that the training data points fall. Afterwards, the frequency estimated is used for inference. However, two approaches have very different effects. With the histogram approach, frequencies of bins are estimated in advance. When novel data arrive, the inference process is conducted without seeing the training data again. With the nearest neighbour approach, an inference process is made only when all the training data points are present. Moreover, various distance metrics are used, for instance the Euclidean distance, the binary distance and other biologically relevant distance measures. Figure 3.2 shows five estimated probability density functions based on five different bin sizes using the nearest neighbour approach. In the Figure, it can be seen that when the number of bins is small, hence large bin size, the estimated density function has a trend to smooth out the peaks. However, when the number of bins is large, hence small bin size, the estimated function has many unnecessary sparks. There is therefore a procedure for determining if the bin size is optimal.

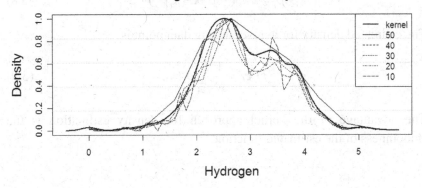

Fig. 3.2. Five estimated probability density functions for the hydrogen data based on five bin sizes using the nearest neighbour approach. Five bin sizes are made by dividing the data interval over the number of bins, e.g. 10, 20, 30, 40, and 50. These five density functions are normalised together with the density function estimated using the kernel approach.

3.3.2 *Kernel approach*

The kernel approach is a kernel learning method and is an extension to the nearest neighbour approach. The kernel approach is also referred to as the Parzen window approach [156]. With the nearest neighbour approach, the frequency of training data points within an interval of a testing data point with a predefined bin size is calculated. This means that all the training data points within the interval play the same role for density estimation. With the kernel approach, such equal weighting is changed. Below is a brief description of the kernel approach for density estimation.

We still use $\Omega = \left\{ \mathbf{x}_n \right\}_{n=1}^{N}$ to denote a training data set. A similarity vector for a novel testing data point is defined as $\left\{ \rho_i \right\}_{i=1}^{N}$, where ρ_i is defined as the similarity between the testing data point and a training data point, namely $\mathbf{x}$ and $\mathbf{x}_i$. The similarity function is named as a kernel function using the kernel method. The commonly used kernel function is the radial-basis function defined as below

$$\rho_i = \exp(-\beta \left\| \mathbf{x} - \mathbf{x}_i \right\|^2) \qquad (3.8)$$

The estimated density for a novel testing data point is

$$f(\mathbf{x}) = \frac{1}{N} \sum_{i=1}^{N} \rho_i(\mathbf{x}) \qquad (3.9)$$

The advantage of the kernel approach for density estimation is the smoothness of the estimated function.

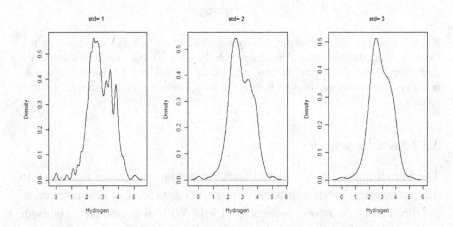

Fig. 3.3. Three estimated probability density functions for the hydrogen in compounds.

Figure 3.3 shows the estimated probability density function using the kernel approach for the hydrogen distribution in compounds. Three values corresponding to the bandwidth are used. In estimating 1-D density function using R, the function is called "density" in which the bandwidth is replaced by the number of standard deviations. Hence three values are one standard deviation (std = 1), two standard deviations (std = 2) and three standard deviations (std = 3). When the bandwidth is larger, the estimated density function is smoother.) For instance, the right panel in Fig. 3.3 shows that some small peaks occurring in the left panel in Fig. 3.3 have been smoothed out.

Shown in Fig. 3.4 are three estimated probability density functions for hydrogen and carbon distributions in compounds. The estimation of 2-D probability density function using R is implemented by the bked2D function. In this function, the bandwidth is explicitly specified. Three values are used for the bandwidth in Fig. 3.4. They are 0.1, 0.5, and 1. The graphs show again that when the bandwidth value is large, the estimated density function smoothes out some peaks which occur in density functions of a small bandwidth value.

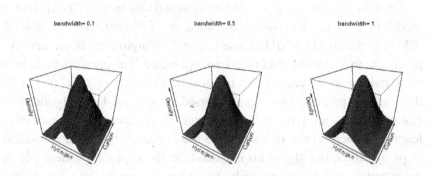

Fig. 3.4. Three estimated probability density functions for the hydrogen and carbon in compounds using the kernel approach.

The semi-parametric approach is an approach between parametric and non-parametric approaches. The estimation process is based on an assumption that data are generated from a model with a number of Gaussians. The model is defined as below

$$f(\mathbf{x}) = \sum_{m=1}^{M} w_m G_m(\mathbf{x}) \tag{3.10}$$

where w_m is the contributing factor (a parameter under estimation) for the mth component, $f(\mathbf{x})$ is the estimated density function, and $G_m(\mathbf{x})$ is the mth component (Gaussian) which is defined as

$$G_m(\mathbf{x}) = \frac{1}{(2\pi)^{d/2}\sqrt{|\Sigma_m|}} \exp\left(-\frac{(\mathbf{x}_i - \boldsymbol{\mu}_m)^{\mathrm{T}} \Sigma_m^{-1} (\mathbf{x}_i - \boldsymbol{\mu}_m)}{2}\right) \quad (3.11)$$

with a mean vector $\boldsymbol{\mu}_m$ and a covariance matrix Σ_m. The contributing factors satisfy two conditions. First, all are positive, $0 \le w_m \le 1$. Second, the sum of them is one

$$\sum_{m=1}^{M} w_m = 1 \quad (3.12)$$

Such an approach is also referred to as mixture models. To fit such a model to a given data set, a so-called the Expectation-Maximisation (EM) algorithm [157-161] is used. The EM algorithm is an iterative procedure in which the parameters are optimised. The parameters include w_m's and the mean vectors and the covariance matrices of $G_m(\mathbf{x})$'s. In the learning process, two steps are used in turn until the algorithm is converged, i.e. until there is no change in parameters in consecutive learning cycles. These two steps are the expectation and maximisation steps by which the algorithm is named. In the expectation step, partial membership ($g_m(\mathbf{x}_n)$) of each data point is computed. The partial membership measures how likely it is that a data point belongs to a component. It is defined as below

$$g_m(\mathbf{x}_n) = \frac{w_m G_m(\mathbf{x}_n)}{\sum_{i=1}^{M} w_i G_i(\mathbf{x}_n)} \quad (3.13)$$

Based on these calculated partial memberships, model parameters are re-computed. The contributing factors are calculated by the following equation

$$w_m = \frac{1}{N} \sum_{n=1}^{N} g_m(\mathbf{x}_n) \quad (3.14)$$

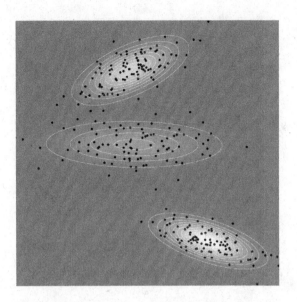

Fig. 3.5. The estimated density function for a data set with three clusters. The estimation is done using the R package "mclust02".

The new mean vectors are re-computed by

$$\boldsymbol{\mu}_m = \sum_{i=1}^{N} \frac{g_m(\mathbf{x}_i)}{\sum_{n=1}^{N} g_m(\mathbf{x}_n)} \mathbf{x}_i \qquad (3.15)$$

If a homogeneous model is the target, the variance is re-computed by

$$\Sigma_m = \sum_{i=1}^{N} \frac{g_m(\mathbf{x}_i)}{\sum_{n=1}^{N} g_m(\mathbf{x}_n)} (\mathbf{x}_i - \boldsymbol{\mu}_m)(\mathbf{x}_i - \boldsymbol{\mu}_m)^{\mathrm{T}} \qquad (3.16)$$

In using the semi-parametric approach, an important issue is to determine the proper number of components, i.e. the value for M. The Bayesian Information Criterion (BIC) [158-162] is one of the commonly used measurements for determining the optimal model structure. By maximising the BIC value, model structure can be optimised. Shown in Fig. 3.5 is the estimated density function for three clusters. Figure 3.6 shows the BIC values for the above case where it shows that the BIC value is maximised at the point with three components ($M=3$).

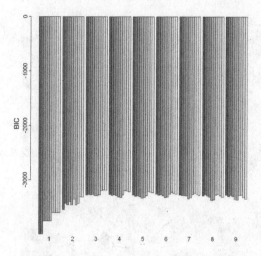

Fig. 3.6. BIC values for the data set with three clusters.

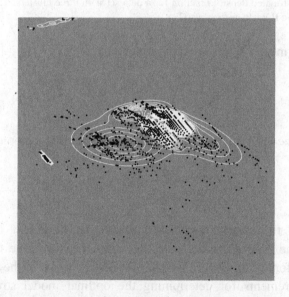

Fig. 3.7. The estimated density function for the compound data using the semi-parametric approach (implemented in the R package by mclust02). The data set is in http://ecsb.ex.ac.uk/book/compoundData.

Figure 3.7 shows an application of the semi-parametric approach to the estimation of density function of compound data in which each compound is coded using nine chemical elements. In order to visualise the density estimation result, data are first mapped to a two-dimensional space using a multi-dimensional scaling approach. The estimation of the density function is conducted in the mapped two-dimensional space. In the estimated density function, it can be seen that there are two densely distributed clusters in the middle while two small clusters are located far away from centre with some loosely distributed compounds.

The second application of this density function estimation is for analysing the impact of acetyllysine on disease development. 453 acetyllysine are collected from the NCBI database. Among them, 40 relate to disease development Ten flanking residues of each acetyllysine are coded using the Kyte-Doolittle hydrophobicity scale [163]. Multi-dimensional scaling is also used before applying the semi-parametric approach for density estimation. Figure 3.8 shows the density estimation where it can be seen that four clusters are formed. The one on the right-hand side corresponds to the 40 acetyllysine involved in disease development.

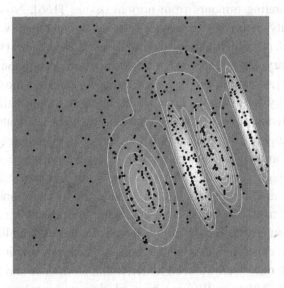

Fig. 3.8. The estimated density function for the acetyllysine data set. The data are in http://ecsb.ex.ac.uk/book/acetyllysine.

Summary

Three major density function estimation approaches have been discussed. Density estimation approaches are able to provide a platform for preliminarily studying data structure prior to implementing a machine learning algorithm to model the data. Because of this, density estimation approaches have been widely employed in various bioinformatics projects. In predicting secondary structure, a novel kernel function was proposed by extending the variance to include information about the distance between a query sequence and the training sequences [122]. The same kernel function has also been used for constructing predictors for species-specific microRNA precursors [164]. For identifying differentially expressed genes between disease-free and disease-related patients, density functions were estimated for making predictive models for disease diagnosis [165]. In analysing the gene expression data of 1536 genes in 100 colorectal cancer and 11 normal tissues, a non-parametric density estimation approach called the iterative local Gaussian clustering (ILGC) was used to identify clusters of genes. The results were similar to those of a semi-parametric approach with three clusters separating tumours from normal tissues [166]. Non-parametric kernel density estimation approach was also combined with entropy approach for selecting highly differentially expressed genes [167].

In summary, the parametric density estimation is commonly a weaker approach for estimating biological data density because most biological data are hardly following a Gaussian distribution. Abnormality is a very common phenomenon. A non-parametric density estimator like the K-nearest neighbour approach and the kernel approach are flexible in constructing unknown density functions. However, such a method has a problem in high computational cost. The semi-parametric approach is based on the assumption that data are generated from a number of basic parametric density functions. Compared with non-parametric approaches, the semi-parametric approach enjoys the advantages of simplicity and the capability of clustering data at the same time. However, it also brings about a challenge when we need to determine an optimal model structure for a data set. Although BIC can be used, data with noise may lead to a difficult situation for selecting an optimal model structure. A better strategy is to use multiple density estimators to investigate the emerged property or data structure.

Chapter 4

Dimension Reduction

Dimension reduction is a technique widely used in many applications. The main objective is to reveal data structure which is hard to obtain from a high-dimensional space through mapping the high-dimensional space to a low-dimensional space. The mapping is commonly conducted by a machine learning algorithm which uses various metrics and various learning strategies. After learning, the new space is commonly 2-dimensional (or 3-dimensional); working in this space we can study how data are clustered and how clusters are mutually correlated. This then provides a basis for further studies including classification analysis, knowledge extraction and hypothesis generation. In this chapter, we discuss two basic dimension reduction algorithms. Importantly, their difference and strengths will be emphasised. The applications to bioinformatics projects are demonstrated as well.

4.1 General

Dimension reduction is a popular topic in machine learning and bioinformatics. It is to find a proper algorithm by which a multi-dimensional data space can be mapped to a low-dimensional space with as small deviation as possible for better visualisation. Denoted by $\mathcal{D} = \{ \mathbf{x}_n \in \Re^d \}_{n=1}^N$ where N is the total number of data points and, $d > 2$ (or $d > 3$) is the dimensionality of $\mathcal{D}$, a machine learning algorithm is used to map $\mathcal{D}$ to $\tilde{\mathcal{D}} = \{ \mathbf{y}_n \in \Re^{\tilde{d}} \}_{n=1}^N$ where $\tilde{d} = 2$ (or $\tilde{d} = 3$) is the dimensionality of $\tilde{\mathcal{D}}$. The process is one-to-one mapping-based, i.e. $\phi : \mathbf{x}_n \mapsto \mathbf{y}_n, \forall n \in [1, N]$. This means that for any original data point $\mathbf{x}_n$ in $\mathcal{D}$ we can find its mapping $\mathbf{y}_n$ in $\tilde{\mathcal{D}}$. Importantly, if we find

38

a nearest neighbour of $\mathbf{x}_n$ (denoted by $\mathbf{x}_m$) in $\mathcal{D}$, we are expected to find $\mathbf{y}_m$ in $\tilde{\mathcal{D}}$ which satisfies

$$\| \mathbf{y}_m - \mathbf{y}_n \| \le \| \mathbf{y}_i - \mathbf{y}_n \|, \forall i \in [1, N] \, \& \, i \ne m \qquad (4.1)$$

where $\forall$ reads as "for all" and $\| \mathbf{y}_m - \mathbf{y}_n \|$ means the distance between $\mathbf{y}_m$ and $\mathbf{y}_n$.

Having understood the general principle of dimension reduction, the next important question is how to select a proper algorithm for a specific application. This requires a clear understanding regarding the strengths of different algorithms. In this chapter, two basic algorithms which are commonly used in bioinformatics for dimension reduction are introduced. They are multi-dimensional scaling and principle component analysis.

There are two commonly used principles involved in various applications. The first is to maintain the information (variance) in the original data as much as possible. The second is to preserve the topological structure of the original data space as unchanged as possible during mapping. It must be emphasised that any mapping from a high-dimensional space to a low-dimensional space will lose information. This is because the complexity in a high-dimensional space is normally not expected to be fully embedded into the low-dimensional space. The larger the difference between the original and the new dimensionality, the more information may be lost during mapping. The larger the complexity in the original high-dimensional space, the more information may be lost during mapping. The complexity in dimension reduction algorithms is then the target of minimising the loss of the information in the original data space.

4.2 Principal component analysis

Principal component analysis (PCA) searches for a set of mutually orthogonal bases which form the new coordinates in the new space and projects the original data space to the new data space through a learning transformation [1, 2]. In the new data space, the coordinates are ordered in terms of projected information (variance). The first coordinate has the largest variance while the following coordinates have decreasing variances. A linear transformation of the matrix of the original data, denoted by $\mathbf{X}$, to the matrix in the new data space, denoted by $\mathbf{Y}$, is defined as

$$\mathbf{Y}^T = \mathbf{X}^T \mathbf{W} \tag{4.2}$$

where $\mathbf{W}$ is the transformation matrix and $\mathbf{X}$ is normalised with zero mean, i.e.

$$E[\mathbf{x}_n] = \mathbf{0}, \forall\, n \in [1, N] \tag{4.3}$$

When only one transformation vector is used we reduce the dimensionality to one,

$$y = \mathbf{X}^T \mathbf{w} \tag{4.4}$$

Shown in Fig. 4.1 is a data set in a two-dimensional space. With this data distribution, we study how we can search for a new data space to which this data set can be mapped with the first new coordinate having the richest information or the largest variance (denoted by $\mathbf{u}_1$). Suppose the mapping is made by $\mathbf{u}$

$$y_n = \mathbf{x}_n^T \mathbf{u} \tag{4.5}$$

The variance of $\mathbf{y}_n$ can be measured by

$$y_n^2 = (\mathbf{x}_n^T \mathbf{u})^2 \tag{4.6}$$

The expectation of the variance is expressed as

$$E[y_n^2] = E[(\mathbf{x}_n^T \mathbf{u})^2] = \mathbf{u}^T \mathbf{X}^T \mathbf{X} \mathbf{u} = \mathbf{u}^T \Sigma \mathbf{u} \tag{4.7}$$

where Σ is the co-variance matrix in the original data space. By maximising the variance $E[\mathbf{y}_n^2]$ we obtain the optimal mapping direction $\mathbf{u}$. $E[\mathbf{y}_n^2]$ is also called the eigen value (λ). When extending the mapping dimensions to $\tilde{d} \leq d$ with new mutually orthogonal coordinates we have

$$\mathbf{U}^T\Sigma\mathbf{U} = \Lambda = \operatorname{diag}\{\lambda_i\}_{i=1}^{\tilde{d}} \qquad (4.8)$$

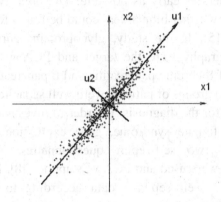

Fig. 4.1. Illustration of PCA. The coordinates (x1, x2) represent the original data space while the coordinates (u1, u2) represent the new data space. The dots are the data points.

In Bishop's book [3], an analysis shows that the loss of the information in the original data space during a dimension reduction process is

$$\sum_{i=\tilde{d}+1}^{d} \lambda_i \qquad (4.9)$$

PCA has been widely used in bioinformatics applications, for instance lactation gene network analysis [4], classification of normal, chronic pancreatitis and pancreatic cancer sera [5], the analysis of low-dose radiation-associated changes in cytokine gene expression profiles [6], the analysis of nucleoside electrophoretic profiles [7], the analysis of chronic fatigue Syndrome through gene expression profile study [8], the study of LC/MS/MS data [9], the analysis of human MHC supertypes [10], the analysis of vitamin E deficiency and metabolic deficits in neuronal ceroid lipofuscinosis using NMR spectroscopy data [11], the analysis of DNA string motifs [12].

In classifying normal, chronic pancreatitis and pancreatic cancer sera, the conventional clinical markers were less accurate and it was desirable to detect the cancers as early as possible. The objective of using PCA was to investigate whether biomarkers could be found for the early-stage cancer detection [5]. In the study, glycoproteins enrichsed by lectin affinity chromatography were the target and PCA was applied to the microarray data of 8 chronic pancreatitis and 6 pancreatic cancer sera. It was found that two groups of patients were well separated thus providing better biomarkers for the diagnosis. In order to investigate the molecular basis of chronic fatigue syndrome, gene expression profiles of 167 participants with two self-report questionnaires (multidimensional fatigue inventory) were used and PCA was applied [8]. It was found that PCA was able to well separate data according to their biological classifications.

Two important aspects must be noted. First, PCA is a linear approach as described in equation (4.2). The nonlinear data structure will not be well-explored. Second, the loss of the information of the original data space using PCA follows equation (4.9). It is very important to check if such a loss is affordable.

4.3 An application of PCA

The data used here has been published in an earlier study [13] and can be seen in the website: http://ecsb.ex.ac.uk/pseudomallei. It was obtained using a proteome array chip to measure antibody responses to a panel of 214 immunoreactive antigens.

Sera from melioidosis positive or negative patients in Singapore were generally taken on admission to hospital or obtained from walk-in clinics. Positive samples (n=87) were taken from patients on admission to hospital and who had a diagnosis of melioidosis confirmed by blood culture. The negative sera (n=59) were taken from patients who were either admitted to hospital or walk-in clinics but were negative for melioidosis.

PCA is applied to this data set to investigate how genes are distributed among the non-infected (negative) and infected (positive) patients. Figure 4.2 shows the PCA for the negative data, where the left panel shows the PCA visualisation using the first two principal components while the right panel shows the distributions of eigen values (variances) across the principal components. The loss of the information in the original data space is 81% using equation (4.9). This means that the knowledge displayed using PCA for this high-dimensional space is less than 20%. In this two-dimensional space, it can be seen that a few genes are distributed in low density areas.

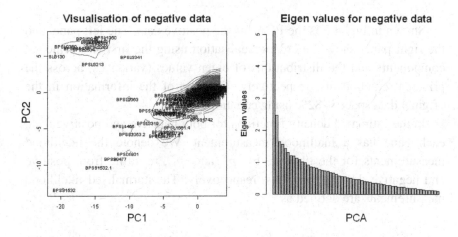

Fig. 4.2. PCA of the negative *Burkholderia pseudomallei* data. The left panel shows the visualisation using two top principal components. The right panel shows the eigen value distribution. The contours represent the density estimated using a kernel approach.

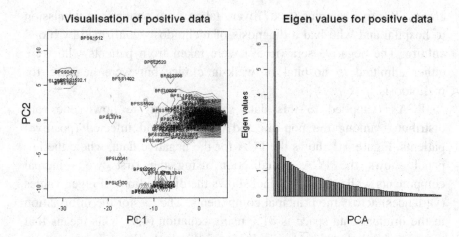

Fig. 4.3. PCA of the positive *Burkholderia pseudomallei* data. The left panel shows the visualisation using two top principal components. The right panel shows the eigen value distribution. The contours represent the density estimated using a kernel approach.

Shown in Fig. 4.3 is the PCA for the positive data. The left panel and the right panel show the PCA visualisation using the first two principal components and the distributions of eigen values (variances) across the principal components, respectively. The loss of the information in the original data space is 82% using equation (4.9).

In the estimated density functions for both negative and positive data, each gene has a likelihood measurement. We denote the likelihood measurements for the ith gene by p_i^+ and p_i^-, resulting from positive and negative density functions, respectively. The normalised likelihood measurements are defined as

$$\tilde{p}_i^+ = \frac{p_i^+}{\sum_i p_i^+} \qquad \tilde{p}_i^- = \frac{p_i^-}{\sum_i p_i^-} \tag{4.10}$$

A PCA differential score can be defined as

$$\lambda_i = \left| \tilde{p}_i^+ - \tilde{p}_i^- \right| \tag{4.11}$$

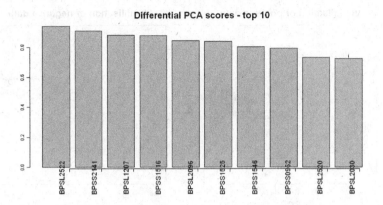

Fig. 4.4. The differential PCA scores for the top ten genes identified using PCA and a kernel density estimation approach. The horizontal axis lists the top ten genes and the vertical axis represents the differential PCA scores.

The larger the differential score, the larger the differential activity in negative and positive PCA models. This means that the top gene which can be regarded as a differential gene using differential PCA scores can be defined as

$$m = \arg\max \left\{ \lambda_i \right\}, \forall i \in [1, d] \tag{4.12}$$

with d as the number of genes (data dimensions). Figure 4.4 shows the differential PCA scores for this data set, where the gene BPSL2522 has been identified as the most differential gene using PCA. The gene has been identified as one of the biomarkers in using genetic programming approach to identify biomarkers for the disease [13].

Shown in Fig. 4.5 are the localisations of the top five genes distributed in both negative (the right panel) and positive (the left panel) PCA maps. It can be seen that these top five genes are located in high density regions of the negative PCA map, but in low density regions of the positive PCA map. Having understood that genes with a large expression measurement normally have a low density measurement,

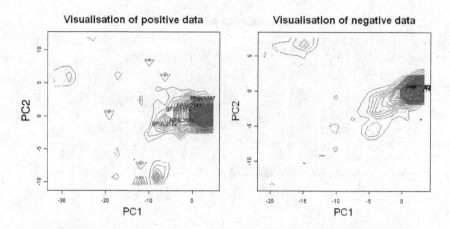

Fig. 4.5. The localisation of the top five genes with high differential PCA scores.

this shows that these top five differential genes demonstrate positive differentiation between positive and negative patients.

4.4 Multi-dimensional scaling

Multi-dimensional scaling aims to visualise high-dimensional data through a learning process which can preserve as much original data structure as possible. Among various multi-dimensional scaling algorithms, the Sammon mapping is the most powerful one. The Sammon mapping was proposed in 1969 by Sammon Jr [14]. The basic principle of the algorithm is to maintain the relative topological structure as unchanged as possible during mapping a high-dimension data space to a low-dimension data space. The relative topological structure is quantified by the pair-wise distance between data points. For N data points, there are $N (N - 1) / 2$ pair-wise distances. Like most other multi-dimensional scaling algorithms, the Sammon mapping also makes one-to-one mapping, i.e. all the original data points can find their locations in a mapping space. There are therefore another set of pair-wise mapping distances between data points. Sammon's idea was to minimise the deviation between the original pair-wise distances and the mapping pair-wise distances. In terms of this, an objective function considers the distance of distances as below

$$\sum_{i=1}^{N} \sum_{j=i+1}^{N} (d_{ij}^* - d_{ij})^2 \tag{4.13}$$

Because it is unavoidable to lose the information in the original data space during mapping from a space with a higher dimensionality to a space with a lower dimensionality, the distance of distances defined in equation (4.13) is merely practical. Sammon proposed the concept of the relative distance, i.e. the normalised distance of distances. The error (objective) function proposed by Sammon was defined as below

$$\mathcal{E} = \frac{1}{\sum_{i<j} d_{ij}^*} \sum_{i=1}^{N} \sum_{j=i+1}^{N} \frac{(d_{ij}^* - d_{ij})^2}{d_{ij}^*} \tag{4.14}$$

A learning process then minimises this error function.

The Sammon mapping algorithm has been used for visualising gene expression data [15-18]. Most applications have indicated that it is better than a linear approach for data visualisation. This results from the nonlinearity property of the Sammon mapping algorithm.

Shown in Fig. 4.6 are examples of using PCA and the Sammon algorithm. On the left panel of Fig. 4.6 (a), the original data is composed of two rings in a two-dimensional space with an added noise of Gaussian $\mathcal{N}(0, 0.1)$ in the third dimension. The PCA map is shown on the middle panel of Fig. 4.6 (a) and the Sammon map is shown on the right panel. It can be seen that the Sammon map can preserve the original data structure well while this data structure can hardly be seen in the PCA map. Figure 4.6 (b) shows another case where PCA fails to preserve the original data structure after mapping. The original data is a sin function with an added noise of Gaussian $\mathcal{N}(0, 0.1)$. The data structure disappears in the PCA map (the middle panel of Fig. 4.6 (b)) while it is well preserved in the Sammon map (the right panel of Fig. 4.6 (b)). These two examples are consistent with some research, for instance, in analysing gene expression data, it has been argued that PCA may not be able to achieve a useful picture unless specific data pre-process is conducted in advance. However, multi-dimensional scaling can deliver better results [19].

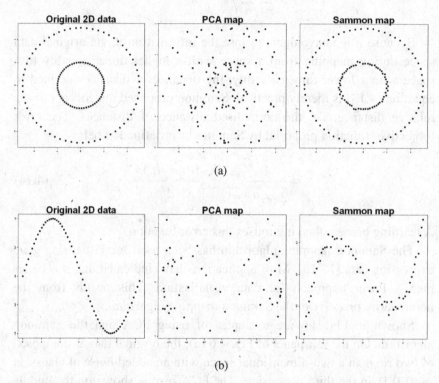

Fig. 4.6. Two comparisons between PCA and the Sammon algorithm. The discussion can be seen in the main text.

4.5 Application of the Sammon algorithm to gene data

We now use the same data used in PCA in this chapter to see if a different pattern is seen using the Sammon mapping algorithm. The data description has been given in the section above. Figure 4.7 shows the Sammon mapping results for both negative and positive *Burkholderia pseudomallei* gene expression data. It can be seen that there is a large difference between the negative and positive maps. Compared with the PCA maps, the Sammon maps show an even larger difference. With the Sammon algorithm, two dimensions are used for the mapping. The positive map displays more genes with low density while the negative map shows more dense distribution.

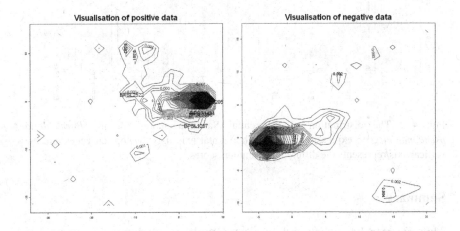

Fig. 4.7. The Sammon mapping results of the *Burkholderia pseudomallei* gene expression data. The left panel shows the map of the positive data while the right panel shows the map of the negative data. Note that unlike PCA, the coordinates have no physical meanings.

We can then use the same approach mentioned in the PCA model constructed for this data set to find top differential genes by estimating density functions for positive and negative Sammon maps. The density functions are estimated for the two coordinates in both Sammon maps using the kernel approach. Differential Sammon scores are derived in the same way as the differential PCA scores mentioned above. Figure 4.8 shows the differential Sammon scores for the top ten genes. The localisations of the top five differential genes are illustrated in Fig. 4.7, where it can be seen that these five genes are positively differentiated.

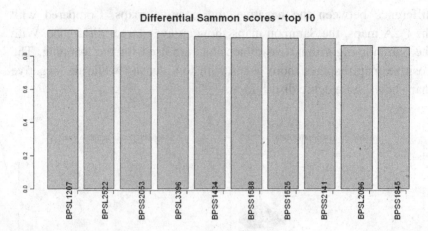

Fig. 4.8. The distribution of differential Sammon scores for the *Burkhoderia pseudomallei* gene expression data. The horizontal axis lists the top ten genes and the vertical axis represents the differential Sammon scores.

Summary

This chapter has discussed two commonly used dimension reduction and visualisation approaches, namely principal component analysis and the Sammon mapping algorithm. They belong to two different statistical machine learning mechanisms. The former is a linear approach while the latter is a nonlinear approach. The former is for preserving the largest variance in data during mapping while the latter is for preserving the topological structure as much as possible during mapping.

It must be noted that PCA is a parameterised system where data structure is learned and maintained in model parameters. For instance, the first principal component will gain the largest variance in data. If a new datum is generated, it is easy to recall its relationship with all the original data without any further learning. However, the Sammon mapping algorithm is not designed for associative memory. Except for the map used for visualisation, there is no way to recall the relationship between a novel datum and the original data.

The new study of PCA has led to two powerful dimension reduction and visualisation approaches. They are probabilistic PCA [20, 21] and nonlinear PCA [22]. They are beyond the scope of this book. Readers can refer to relevant articles for details.

Chapter 5

Cluster Analysis

This chapter focuses on one fundamental issue in analysing biological data, i.e. how to find scientific laws which are hidden in data. Grouping and partitioning data are two very powerful approaches for discovering relevant biological regulations which can then be used in late hypothesis verification. In machine learning, such an approach is called cluster analysis, a type of unsupervised learning approach. The grouping data approach puts the emphasis on data relationship re-construction i.e. exploring how data are clustered through a learning process. Partitioning data, on the other hand, is to discover hidden data structure through a learning process. Compared with the grouping data approach, the partitioning data approach puts the emphasis on a comprehensive data structure and the predictive capability of the discovered data structure. In this chapter, four fundamental clustering algorithms are introduced and their applications to bioinformatics are demonstrated. The four algorithms are the hierarchical clustering approach, the K-means algorithm, the fuzzy C-means algorithm, and the mixture models.

5.1 Hierarchical clustering

The hierarchical clustering approach is a grouping data approach, where the aim is to build a relational and hierarchical structure to explore and represent mutual relationships between data points. The basis is to find related data points and then group them rationally for interpreting data or making biological hypotheses. The basic technique for interpreting mutual relationship between data points is correlation analysis (or similarity calculation). If two d-dimensional vectors are denoted by

$\mathbf{x}_n \in \mathfrak{R}^d$ and $\mathbf{x}_m \in \mathfrak{R}^d$, the dissimilarity (distance) between them is defined as

$$d(\mathbf{x}_n, \mathbf{x}_m) = \| \mathbf{x}_n - \mathbf{x}_m \| = \sum_{i=1}^{d} (x_{ni} - x_{mi})^2 \tag{5.1}$$

where x_{ni} and x_{mi} are the ith elements of $\mathbf{x}_n$ and $\mathbf{x}_m$ respectively. The Euclidean distance is used in equation (5.1), but metrics can be used in different applications.

During a simple hierarchical clustering, a pair of data points with the highest similarity is grouped or merged. This process is progressive until one cluster is formed, i.e. all data points are in one super cluster. For $\mathcal{D}^N = \{ \mathbf{x}_1, \mathbf{x}_2, \cdots, \mathbf{x}_N \}$, the first sub-cluster is formed for $\mathbf{x}_i$ and $\mathbf{x}_j$ if

$$d(\mathbf{x}_i, \mathbf{x}_j) = \min\{ d(\mathbf{x}_n, \mathbf{x}_m) \}, \forall \mathbf{x}_n, \mathbf{x}_m \in \mathcal{D}^N \tag{5.2}$$

A mean vector for $\mathbf{x}_i$ and $\mathbf{x}_j$ is calculated and is expressed as $\boldsymbol{\mu}_1$. This mean vector is added into the data set while $\mathbf{x}_i$ and $\mathbf{x}_j$ are removed from the data set

$$\mathcal{D}^{N+1} = (\mathcal{D}^N - \{ \mathbf{x}_i, \mathbf{x}_j \}) \bigcup \{ \boldsymbol{\mu}_1 \} \tag{5.3}$$

In the next step, an original data point $\mathbf{x}_i$ may have the smallest distance with an original data point $\tilde{\mathbf{x}}_j = \mathbf{x}_j \in \mathcal{D}^{N+1}$ or a mean vector of the sub-cluster $\tilde{\mathbf{x}}_j = \boldsymbol{\mu}_1 \in \mathcal{D}^{N+1}$

$$d(\mathbf{x}_i, \tilde{\mathbf{x}}_j) = \min\{ d(\mathbf{x}_n, \tilde{\mathbf{x}}_m) \}, \forall \mathbf{x}_n, \tilde{\mathbf{x}}_m \in \mathcal{D}^{N+1} \tag{5.4}$$

This merge generates the second sub-cluster as well as its mean vector $\boldsymbol{\mu}_2$. The data set is updated as below

$$\mathcal{D}^{N+2} = (\mathcal{D}^{N+1} - \{ \mathbf{x}_i, \tilde{\mathbf{x}}_j \}) \bigcup \{ \boldsymbol{\mu}_2 \} \tag{5.5}$$

It can be seen that $\left| \mathcal{D}^{N+k} \right| < \left| \mathcal{D}^{N+k-1} \right|$. If $\left| \mathcal{D}^{N+k} \right| > 1$, merging continues

$$d\,(\tilde{\mathbf{x}}_i, \tilde{\mathbf{x}}_j) = \min\{d\,(\tilde{\mathbf{x}}_n, \tilde{\mathbf{x}}_m)\}, \forall \tilde{\mathbf{x}}_n, \tilde{\mathbf{x}}_m \in \mathcal{D}^{N+K} \qquad (5.6)$$

The merging continues until $\left| \mathcal{D}^{N+K} \right| = 1$.

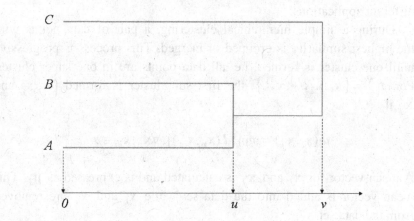

Fig. 5.1. An illustration of the hierarchical clustering approach.

The hierarchical clustering approach has two distinct features which may not be seen in other clustering algorithms. First, relationship between data points can be well visualised. Second, the merging distance can be well used for interpreting data. Figure 5.1 shows an example where the cluster distance is u for data points A and B. The cluster distance is increased to $u > v$ to include data point C.

In bioinformatics, the hierarchical clustering approach has been used for identifying protein relationships based on spectral properties [186], diagnosing chronic fatigue Syndrome based on gene expression profile [172], and detecting esophageal cancer using gene expression profile [187].

5.2 K-means

Rather than aiming to explore the relationship between data points of a data set, the K-means algorithm [158] is for learning how data are structured or investigating the data structure from which data are generated. Using the same notation of data vectors as mentioned above, the K-means algorithm assumes that data are generated from K clusters, hence it tries to partition data into these K clusters with the smallest diversity

$$\min\left\{ \sum_k \sum_{\mathbf{x}_n} \| \mathbf{x}_n - \boldsymbol{\mu}_k \|^2 \right\}, \forall \mathbf{x}_n \in \vartheta_k \qquad (5.7)$$

The centre of the kth cluster is defined as

$$\boldsymbol{\mu}_k = \frac{1}{|\vartheta_k|} \sum_{\mathbf{x}_n} \mathbf{x}_n, \forall \mathbf{x}_n \in \vartheta_k \qquad (5.8)$$

where $|\vartheta_k|$ is the number of data points in the kth cluster.

In order to find centres of K clusters, we need to start with K centres which are random values. Based on the random centres, each data point is assigned to a cluster by

$$m = \arg\min\{ \| \mathbf{x}_n - \boldsymbol{\mu}_k \|^2 \} \Rightarrow \mathbf{x}_n \mapsto \vartheta_m, \forall k, m \in [1, K] \qquad (5.9)$$

Here $\mathbf{x}_n \mapsto \vartheta_m$ means that $\mathbf{x}_n$ has been mapped or assigned to ϑ_m because $\mathbf{x}_n$ has the smallest distance with the kth cluster. After this, K centres are updated using equation (5.8) and a new assignment process is carried out. The learning process will continue until K centres are stable, i.e. the updated centres in two consecutive learning cycles have no or little change.

Using the K-means algorithm, one difficult issue is how to determine an accurate cluster structure, i.e. the number K. This is a common problem of model selection in machine learning. For instance, for a 4-cluster data structure, different guesses of K will lead to different cluster structure shown in Fig. 5.2.

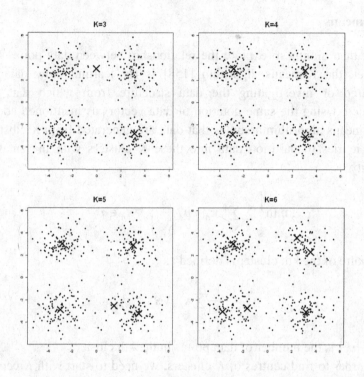

Fig. 5.2. Four cluster structures with four different guesses of K using the K-means algorithm. The dots are the original data points and crosses are the estimated cluster centres. The data are generated from four Gaussian distributions with the centres as (-3, -3), (-3, 3), (3, -3), and (3, 3). The standard deviation is one.

In order to estimate the right cluster structure for a data set, we have to introduce a parameter, such as the difference of the within-cluster diversity. For the simple case mentioned above, the within-cluster diversity can be informative. For instance, the within- cluster diversities of the case mentioned in Fig. 5.2 using different guessed cluster numbers are seen in Fig. 5.3. It can be seen that the difference of the within-cluster diversity is minimised when the guessed cluster number is either two or four. However, this measure is getting confused if the cluster number should be two or four because both of these two structures have the smallest deviations.

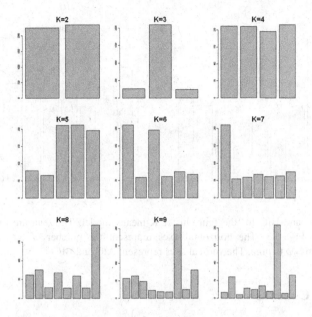

Fig. 5.3. Within-cluster diversity of the example mentioned in Fig. 5.2 using different guessed cluster numbers from two to nine. The horizontal axes represent the number of guessed clusters. The vertical axes represent the within-cluster diversity.

A number of statistical measures have been proposed for model selection problems, for instance, the Akaike Information Criterion (AIC) and Schwarz's Bayesian information criterion (BIC) [162].

$$-2\log \mathcal{L} + \lambda M \qquad (5.10)$$

where $\mathcal{L}$ is the model likelihood, M is the number of model parameters and λ is a constant. $\lambda = 2$ for AIC and $\lambda = \log N$ for BIC, where N is the number of data points. Both criteria are minimised to select the best model. To use these two criteria, model likelihood must be provided. In a K-means cluster model, the likelihood can be calculated by assuming data points in each cluster follow a multivariate Gaussain distribution. Figure 5.4 shows AIC and BIC for the four-cluster K-means model discussed above. It can be seen that both AIC and BIC have successfully detected the cluster structure accurately, i.e. both criteria have been minimised at the number four.

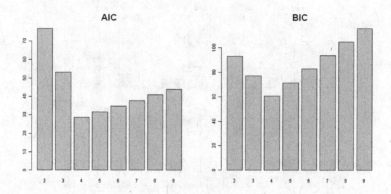

Fig. 5.4. AIC and BIC for the four-cluster K-means model. The data are the same as that used in Fig. 5.3. The horizontal axes represent the number of guessed cluster numbers from *two* to *nine*. The vertical axes represent AIC and BIC.

5.3 Fuzzy C-means

The fuzzy c-means algorithm was proposed in 1981 [188, 189]. The centre of a cluster defined in equation (5.8) is changed to

$$\boldsymbol{\mu}_k = \sum_{\mathbf{x}_n} f(\mathbf{x}_n)\,\mathbf{x}_n, \forall\, \mathbf{x}_n \in \vartheta_k \qquad (5.11)$$

If

$$f_k(\mathbf{x}_n) = \frac{1}{|\vartheta_k|} \qquad (5.12)$$

Each data point in a cluster plays an equal role (membership) in forming a cluster. However, the *soft* membership used in the fuzzy c-means algorithm is more realistic, i.e. $f_k(\mathbf{x}_n)$ is not a constant within a cluster in some applications. In the algorithm, the objective function is defined as

$$\jmath = \sum_{n=1}^{N} \sum_{k=1}^{K} \left[f_k(\mathbf{x}_n) \right]^m \left\| \mathbf{x}_n - \boldsymbol{\mu}_k \right\|^2, \forall\, m \in [1, \infty] \qquad (5.13)$$

where μ_k is the centre of the kth cluster and $f_k(\mathbf{x}_n) \in [0,1]$ is the membership that $\mathbf{x}_n$ belongs to the kth cluster. The m parameters are used to weight the memberships. The centres are defined as below

$$\mu_k = \frac{\sum_{n=1}^{N} [f_k(\mathbf{x}_n)]^m \mathbf{x}_n}{\sum_{n=1}^{N} [f_k(\mathbf{x}_n)]^m} \qquad (5.14)$$

while the membership is defined as below

$$f_j(\mathbf{x}_n) = \sum_{k=1}^{K} \left(\frac{\|\mathbf{x}_n - \mu_j\|}{\|\mathbf{x}_n - \mu_k\|} \right)^{-\frac{2}{m-1}} \qquad (5.15)$$

The algorithm starts from random guesses for the centres as the K-means algorithm. Based on the guessed centres, memberships are estimated. This is similar to the K-means algorithm where the distance between a data point and the centre of a cluster is used to determine if the data point belongs to the cluster. Based on the calculated membership values, new centres are calculated using equation (5.14). These two calculations are repeated until maximum cycles are reached or the centres do not change much.

Compared with the K-means algorithm which uses a *hard* membership function, the fuzzy c-means algorithm benefits from its continuous hence soft membership function from which the centres of clusters can be more accurately estimated. Figure 5.5 shows a case where four clusters are more overlapping. In this case, the K-means algorithm often wrongly estimates cluster centres (as shown in the left panel of Fig. 5.5) while the fuzzy c-means algorithm is able to consistently estimate correct centres of four clusters.

It must be noted that the fuzzy c-means algorithm is also unable to determine the cluster structure automatically. For the same data used in Fig. 5.3, AIC and BIC clearly indicate that four clusters are the best data structure for the data. AIC and BIC are shown in Fig. 5.6 when using the fuzzy c-means algorithm.

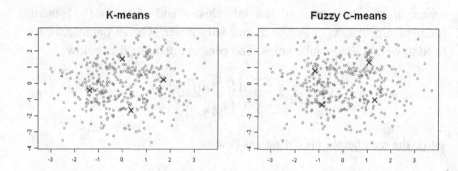

Fig. 5.5. A comparison between the K-means and the fuzzy c-means algorithms. The data are generated from four Gaussian distributions with the centres as (-1, -1), (-1, 1), (1, -1), and (1, 1). The standard deviation is one. Small dots represent the original data and crosses represent cluster centres.

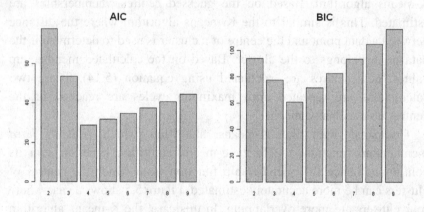

Fig. 5.6. AIC and BIC for the four-cluster fuzzy c-means model. The data are the same as that used in Fig. 5.3. The horizontal axes represent the number of guessed cluster numbers from *two* to *nine*. The vertical axes represent the calculations of AIC and BIC.

5.4 Gaussian mixture models

Both the K-means algorithm and the fuzzy c-means algorithm are designed to detect cluster structure with cluster densities distributed homogenously in all dimensions, i.e. the volume of each cluster is symmetrical with respect to the cluster centre. They therefore have problems in detecting cluster structures with clusters in which different dimensions are correlated. This problem can be well addressed in the mixture model algorithm.

In a mixture model [158, 159, 190-192], the membership function is defined as a probability while the objective function is defined as a likelihood function

$$J = \prod_{n=1}^{N} \sum_{k=1}^{K} w_k p(\mathbf{x}_n \mid k) = \prod_{n=1}^{N} p(\mathbf{x}_n) \qquad (5.16)$$

where $p(\mathbf{x}_n \mid k)$ is the probability that $\mathbf{x}_n \in \mathfrak{R}^d$ belongs to the kth cluster, $w_k \in [0,1]$ is the mixing coefficient of the kth cluster and

$$\sum_{k=1}^{K} w_k = 1 \qquad (5.17)$$

The Gaussian mixture model is a special case of a widely used mixture model. The probability function used in the Gaussian mixture model is defined as

$$p(\mathbf{x}_n \mid k) = \frac{1}{(2\pi)^{d/2} \sqrt{|\Sigma|}} \exp\left(-\frac{(\mathbf{x}_n - \boldsymbol{\mu}_k)^T \Sigma^{-1} (\mathbf{x}_n - \boldsymbol{\mu}_k)}{2}\right) \qquad (5.18)$$

where $\boldsymbol{\mu}_k$ is the centre of the kth cluster and Σ is the covariance matrix of the kth cluster.

The parameters in a mixture model then include the cluster centres, covariance matrix and mixing coefficients. To estimate these parameters, the expectation-maximisation (EM) algorithm [157, 159] is used. The EM algorithm is a two-step iterative procedure for parameter estimation starting from random guesses of the parameters like the K-means and fuzzy c-means algorithms. The two steps are called the E step and the M step. In the E step, the probabilities are calculated based on the current values assigned to the parameters. The calculated probabilities are then used to update the parameters in the M step.

To understand why we need to use the EM algorithm, we make a simple case, i.e. data in a one-dimensional space. We also make the following simplification

$$\beta_k = \frac{1}{2\sigma_k^2} \tag{5.19}$$

Equation (5.18) is then re-written as

$$p(x_n \mid k) = \sqrt{\frac{\beta_k}{\pi}} \exp(-\beta_k (x_n - \mu_k)^2) \tag{5.20}$$

Applying logarithm to the likelihood function and negating it leads to

$$O = -\sum_{n=1}^{N} \log \sum_{k=1}^{K} w_k p(\mathbf{x}_n \mid k) + \lambda \left(\sum_{k=1}^{K} w_k - 1 \right)^2 \tag{5.21}$$

where $\lambda > 0$ is the Lagrange constant. Letting the derivative of O with respect to β_k being zero leads to

$$\frac{1}{\beta_k} = \frac{\displaystyle\sum_{n=1}^{N} \frac{w_k p(x_n \mid k)}{p(x_n)}(x_n - \mu_k)^2}{\displaystyle\sum_{n=1}^{N} \frac{w_k p(x_n \mid k)}{p(x_n)}} \tag{5.22}$$

The equation is not analytically-solvable because the right-hand side of the equation is a function of β_k. The same thing happens to other parameters. Using the EM algorithm, we start with a random guess for β_k, which is denoted by β_k^0. Based on β_k^0 and initial guesses of the other parameters, we can calculate

$$\tau_{n,k}^0 = \frac{w_k^0 p^0(x_n \mid k)}{p^0(x_n)} \tag{5.23}$$

From $\tau_{n,k}^0$, β_k is updated from β_k^0 to β_k^1 using equation (5.22). These two steps are used in turn until parameters are converged or the maximal learning cycles are approached.

We then go back to the original multi-dimensional space. The updated equation for the cluster centre is

$$\mu_k^{t+1} = \frac{\sum_{n=1}^{N} \tau_{n,k}^t \mathbf{x}_n}{\sum_{n=1}^{N} \tau_{n,k}^t} = \sum_{n=1}^{N} \phi_{n,k}^t \mathbf{x}_n = \mathbf{X}^T \varphi_k^t \qquad (5.24)$$

where t is the iteration time and

$$\varphi_k^t = (\phi_{1,k}^t, \phi_{2,k}^t, \cdots, \phi_{N,k}^t) \qquad (5.25)$$

The update equation for the mixing coefficient is

$$w_k^{t+1} = \frac{\sum_{n=1}^{N} \tau_{n,k}^t}{N} \text{ or } \mathbf{w}^{t+1} = (\mathbf{\Gamma}^t)^T \mathbf{i}_{1/N} \qquad (5.26)$$

where $\mathbf{i}_{1/N} = \left(\dfrac{1}{N}, \dfrac{1}{N}, \cdots, \dfrac{1}{N} \right)$ and $\mathbf{\Gamma}^t = \{\tau_{n,k}^t\}$. The updated equation for the covariance matrix is

$$\Sigma_k^t = \sum_{n=1}^{N} \phi_{n,k}^t (\mathbf{x}_n - \mu_k^t)(\mathbf{x}_n - \mu_k^t)^T \qquad (5.27)$$

Figure 5.7 shows a comparison of three clustering algorithms for a data set with three clusters, all with heterogeneous distributions across dimensions. If a data point is correctly classified, it is printed in gray with a smaller fond size. If a data point is mis-classified, it is printed in dark with a larger font size. For instance, two data points labelled by "3" and printed by a larger font size are classified as members of the cluster of data points labelled by "2" using the K-means algorithm. It can be seen that the mixture models algorithm performs the best with no mis-classified data point.

The mixture models algorithm is a probabilistic algorithm. Because of this, the use of AIC and BIC is straightforward and the performance is better than that of the other two algorithms.

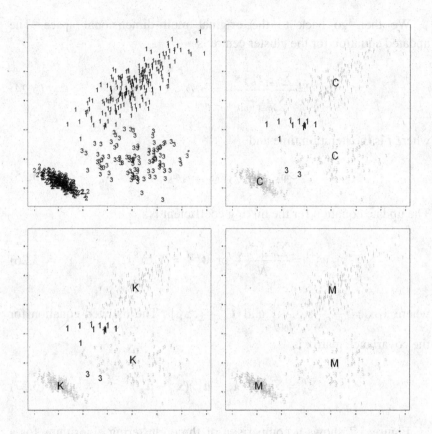

Fig. 5.7. Acomparison between three algorithms for a data set. "K", "C" ands "M" mean the cluster centres found by the K-means, fuzzy c-means mixture models algorithms. The top-left panel shows the original data labels of three clusters.

5.5 Application of clustering algorithms to the *Burkholderiai pseudomallei* gene expression data

We then apply the clustering algorithms to the same *Burkholderia pseudomallei* gene expression data mentioned in the last chapter. The models generated by the hierarchical clustering algorithm for negative and positive data are shown in Fig. 5.8.

Hierarchical cluster analysis - negative data

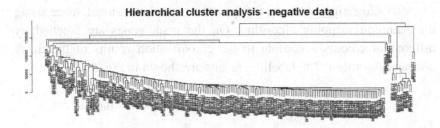

(a) the hierarchical clustering model for the negative data

Hierarchical cluster analysis - positive data

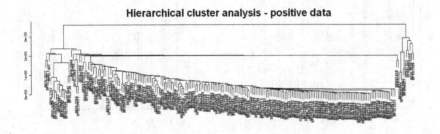

(b) the hierarchical clustering model for the positive data

Fig. 5.8. Clustering models generated using the hierarchical clustering algorithm for the *Burkholderia pseudomallei* gene expression data.

It can be seen from both clustering models that there is a very large cluster and a few small clusters. This is consistent with biological expectation that most genes will not show high differential activity in responding to the external signals while a few will demonstrate differential functions. The genes which do not respond to external signals should have similar expression profiles across samples, hence being clustered together. However, differential genes will have different levels of expressions responding to the external signals, hence demonstrating diversities.

The AIC and BIC calculations using the fuzzy c-means algorithm for the *Burkholderia pseudomallei* gene expression data are shown in Fig. 5.9, where it can be seen that the best cluster numbers are 10 and 12.

After clustering, data are mapped to a two-dimensional space using the Sammon mapping algorithm. On the map, genes are labelled by differential colours according to the classification results of the fuzzy c-means algorithm. The labelling results are shown in Fig. 5.10.

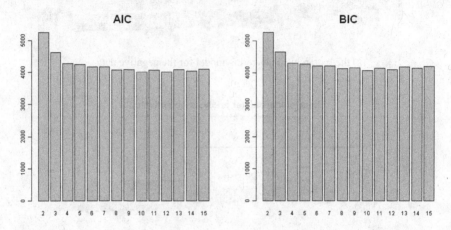

Fig. 5.9. AIC and BIC for the fuzzy c-means model of the *Burkholderia pseudomallei* data. The horizontal axes represent the cluster numbers from two to 15.

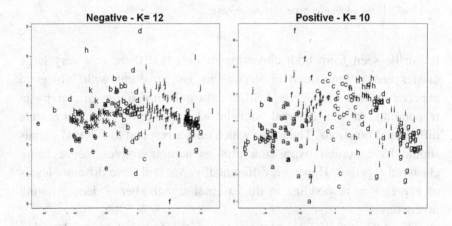

Fig. 5.10. The labelled genes according to the classification results using the fuzzy c-means algorithm for the *Burkholderia pseudomallei* gene expression data. The two maps are generated using the Sammon mapping algorithm. The left panel is for the negative samples and the right panel is for the positive samples.

The Gaussian mixture models algorithm has also been applied to the same gene expression data. The classification results are shown in Fig. 5.11. BIC is used to find the optimal cluster structure. The best cluster numbers are nine and eight for the negative and positive samples respectively.

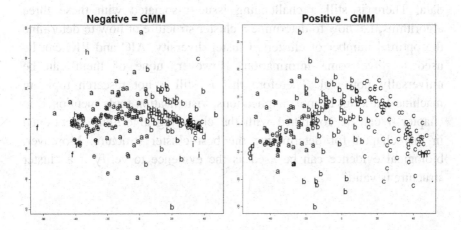

Fig. 5.11. The classification results of the Gaussian mixture models algorithm applied to the *Burkholderia pseudomallei* gene expression data. The two maps are produced using the Sammon mapping algorithm. The negative sample map is in the left panel and the positive sample map is in the right panel.

Summary

This chapter has discussed four basic clustering algorithms. They are the hierarchical clustering algorithm, the K-means, the fuzzy c-means and the mixture models algorithms. Some demonstrations are given to show their differences. In general, the hierarchical clustering algorithm has one typical advantage, i.e. it can visualise the similarity distance between each pair of data points. This is particularly important for explanation research. For instance it can be used to infer evolutionary information from sequence data. However, it does not provide any facility for associative memory. This means that a hierarchical clustering model only supports the interpretation for the existence model. In order to use a built model to interpret unseen data, we have to consider other algorithms such as the aforementioned clustering algorithms like the K-means, the fuzzy c-means and the mixture models algorithms. These algorithms will not provide direct data for visualising data structure. Instead, they can partition data into groups. Within each group, centres are found as typical patterns. The patterns can be used for future inference on unseen data. There is still a challenging issue associated with these three algorithms, i.e. how to determine a cluster structure or how to determine the optimal number of clusters. Cluster diversity, AIC and BIC can be used to give some information. However, none of them can be universally powerful. Therefore this is still a hot research topic in machine learning. In most applications, visualisation tools mentioned in chapter 4 can be combined with the clustering algorithms mentioned in this chapter for determining the best cluster structure. Moreover, biological evidence can be used as the evidence to verify if a cluster structure is valid.

Chapter 6

Self-Organising Map

In the previous two chapters, data reduction (visualisation) and data partitioning algorithms were discussed. A data reduction algorithm may not provide data for data partitioning while a data partitioning algorithm may not visualise data. Self-organising map (SOM) is a neural learning algorithm which is able to combine two categories of algorithms into one system. In this chapter, we study the basic structure and learning rules of SOM. Because SOM has a close relationship to vector quantization, we first introduce vector quantization in this chapter. Demonstrations and case study are also given in relevant places.

6.1 Vector quantization

Vector quantization (VQ) was introduced in the late 1970s and early 1980s [193-196]. VQ is designed for data compression, i.e. representing N data points (input numeric vectors) using M data points (code numeric vectors or representative numeric vectors) where $M<N$. The compression is constrained in a two-dimensional cell-based map. Each cell is associated with a code vector. In this way, the approximate distribution of the input numeric vectors can be visualised. Let's denote $\mathbf{x}_n \in \Re^d$ as the nth d-dimensional input vector and $\mathbf{y}_m \in \Re^d$ as the mth d-dimensional code vector. The collection of all input vectors is denoted by $\mathcal{D}$ and the collection of all code vectors (codebook) is denoted by $\mathcal{C}$. For each input vector, a closest code vector is formed by

$$\| \mathbf{x}_n - \mathbf{y}_m \| = \min\{ \| \mathbf{x}_n - \mathbf{y}_i \| \}, \forall i \in [1,M] \qquad (6.1)$$

This means that the mth code vector is picked from M code vectors to represent the nth input vector through this minimisation process. We use $\phi(\mathbf{x}_n)$ to denote the closest code vector of $\mathbf{x}_n$.

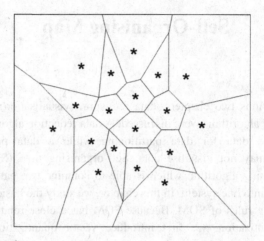

Fig. 6.1. An illustration of VQ. Each cell has a star representing a code vector. Each cell has a number of input vectors with a certain density. The code vector space is composed of 18 code vectors expressed by stars.

In order to find an optimal mapping or optimal distribution visualisation of the input vectors, an objective function is defined as

$$o = \sum_{n=1}^{N} \left\| \mathbf{x}_n - \phi(\mathbf{x}_n) \right\| f(\mathbf{x}_n) \tag{6.2}$$

where $f(\mathbf{x}_n)$ is the density of $\mathbf{x}_n$. A compression process of VQ is to minimise the objective function, i.e. $c^* = \arg\min\{o\}$ where c^* stands for an optimal solution of all possible c s. Figure 6.1 shows the principle of compressing numeric vectors using code vectors. The whole data set of $\mathcal{D}$ is compressed into a smaller space of code vectors, i.e. c. Each cell is denoted by ϑ_m.

In searching for a codebook c with a given number of code vectors, there are two optimality criteria called the nearest neighbour condition

(NC) and the centroid condition (CC). Both must be satisfied. NC requires that each cell must be composed of input vectors satisfying

$$\vartheta_m = \{ \, \mathbf{x} : \| \mathbf{x} - \mathbf{y}_m \| \le \| \mathbf{x} - \mathbf{y}_i \|, \forall i \ne m \} \tag{6.3}$$

The condition $\| \mathbf{x} - \mathbf{y}_m \| \le \| \mathbf{x} - \mathbf{y}_i \|$ implies that if $\mathbf{x}$ is compressed to $\mathbf{y}_m$, $\mathbf{y}_m$ must be $\mathbf{x}$'s nearest neighbour. CC requires each code vector to be the mean vector of all input vectors falling in the cell of the code vector, i.e.

$$\mathbf{y}_m = \frac{\sum_{\mathbf{x}_n \in \vartheta_m} \mathbf{x}_n}{|\vartheta_m|} \tag{6.4}$$

The LBG (Linde, Buzo and Gray) algorithm was proposed to tackle the VQ problem by considering the density function $f(\mathbf{x}_n)$ [194]. The algorithm is an iterative learning procedure. The procedure is shown as bellow

Step 1: initialisation: find the first code vector (M=1) which is the mean vector of all input vectors

$$\mathbf{y}_1 = \frac{1}{N} \sum_{n=1}^{N} \mathbf{x}_n \tag{6.5}$$

The error is calculated as

$$E = \frac{1}{Nd} \sum_{n=1}^{N} \| \mathbf{x}_n - \mathbf{y}_1 \|^2 \tag{6.6}$$

Step 2: splitting: each code vector is split into two given M code vectors. This doubles the number of code vectors as below

$$\begin{aligned} \tilde{\mathbf{y}}_i &= (1 + \varepsilon) \, \mathbf{y}_i \\ \tilde{\mathbf{y}}_{M+i} &= (1 - \varepsilon) \, \mathbf{y}_i \end{aligned} \tag{6.7}$$

where $\varepsilon > 0$ is a small value. A new code book with initialised code vectors is formed with the size increased to $2M$, i.e. $c = \{\mathbf{y}_i\} \bigcap \{\tilde{\mathbf{y}}_i\}$.

Step 3: refining: we first find the nearest neighbour for each input vector, $\phi(\mathbf{x}_n)$. Each code vector in c is updated using equation (6.4). A new error is calculated using equation (6.6). If the old error is denoted by E^0 and the new error is denoted by E^1, an error ratio (π) is calculated by

$$\pi = \frac{E^0 - E^1}{E^0} \tag{6.8}$$

If $\pi > \varepsilon$, a new cycle is repeated to find the new nearest neighbour, to update code vectors and to calculate the error.

Step 2 and step 3 are iterated until the predefined code vector number is approached.

In fact, the core part of step 3 uses the K-means algorithm. A two-dimensional example using two different numbers of code vectors is shown in Fig. 6.2. It can be seen that the code vectors are uniformly distributed in the data area. This is because no density function is used in the LBG algorithm.

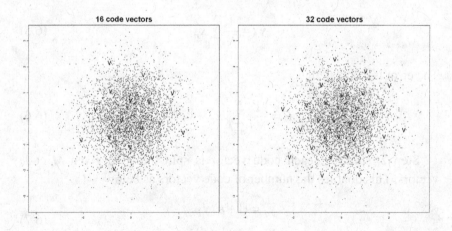

Fig. 6.2. 4000 input vectors are in two-dimensional space in a Gaussian distribution with zero mean and one standard deviation in both dimensions. $\varepsilon = 0.001$. Two panels are generated using the LBG algorithm. The left panel uses 16 code vectors while the right panel uses 32 code vectors.

6.2 SOM structure

SOM was introduced by von der Malsburg [197] and Kohonen [145, 198]. Unlike VQ, self-organising map (SOM) has introduced a number of new features. First, it fixes the cell positions for code vectors in a two-dimensional array. The cell array is shown in Fig. 6.3, where each circle represents a cell which is also referred to as a neuron [145]. The code vector of each neuron is fully connected to the input vectors as shown in Fig. 6.4. The left panel shows a rectangular map while the right panel shows a hexagonal map.

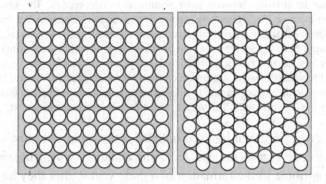

Fig. 6.3. Arrays of cells with ten rows and ten columns. Each cell has a code vector connected to input vectors as shown in Fig. 6.4.

Figure 6.4 shows how data stored in input vectors are used by code vectors of neurons. In this Figure, we consider four input variables, i.e. each input vector has four dimensions. Each code vector also has four dimensions. All neurons are connected to four input variables. An input vector is mapped to a neuron according to the similarity. If an input vector (x_n) has the largest similarity with a code vector (w_m), x_n is mapped to the neuron of w_m. We use w_m rather than y_m to follow the convention of SOM.

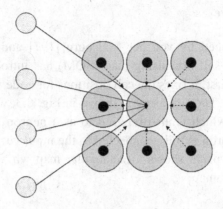

Fig. 6.4. The relationship between input vectors and code vectors. The small circles represent the input vectors. Here there are four input variables meaning that the data are in four-dimensional space. The filled large circles represent cells or neurons. The dots within the large filled circles represent the code vectors (the middle one is omitted). The solid arrows indicate the data flow direction, i.e. data stored in each input vector are fed to the model towards the target neuron which is also referred to as the winner [145]. Here the middle one is the target neuron. The dashed arrows indicate the update directions of code vectors in all nearby cells. The update mechanism is seen in the main text.

The second feature introduced in SOM is the online learning strategy. In VQ, the code vectors are updated using equation (6.4), where all input vectors contribute to the formation of a code vector after they have been confirmed as falling in the cell of the code vector. However, it is no longer used in SOM. Instead, SOM updates code vectors whenever one input vector is fed into the model. The third feature used in SOM is neighbourhood. In VQ, only one code vector is updated using all the input vectors falling into its cell. This means that each input vector only contributes to the update of one code vector. However, this has been changed in SOM. When one input vector is fed to a model, only one neuron is selected as the target neuron whose code vector has the smallest distance with the fed input vector. Centred by this target neuron, a number of neighbouring neurons are determined. Only the code vectors within a neighbourhood are updated. These two features make SOM very different from VQ in that it can preserve the topological structure during learning. In other words, similar input vectors, if they are not mapped to the same neuron (cell), will be mapped to the nearby neurons (cells). The fourth feature implemented in SOM for efficient learning is dynamic learning parameters. Here learning parameters are used to determine learning efficiency, for instance the learning rate which is discussed in the next section. Such an introduction of dynamic learning can avoid possible unnecessary long learning time.

6.3 SOM learning algorithm

Let's denote x_n as the nth input vector and w_m the mth code vector, where $n \in [1, N]$ and $m \in [1, M]$. The relationship (distance) between them is defined as

$$o_{nm} = \frac{1}{2} \| x_n - w_m \|^2 \tag{6.9}$$

A target neuron is selected by minimising this distance

$$\phi(x_n) = \arg\min\{o_{nm}\}, \forall m \in [1, M] \tag{6.10}$$

where $\phi(x_n)$ is the target neuron and its code vector is w_k with $k = \phi(x_n)$. Centred at $\phi(x_n)$, a neighbourhood is formed shown in Fig. 6.5. From this, a set of neurons is formed and is denoted by $\Phi(x_n)$.

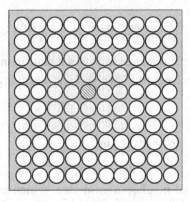

Fig. 6.5. An illustration of neighbourhood of a target neuron. The shaded circle represents the target neuron while the white circles are the neurons within the neighbourhood.

The code vectors in $\Phi(\mathbf{x}_n)$ are updated in magnitude negatively proportional to the amount obtained from differentiation of equation (6.9)

$$\Delta \mathbf{w}_m \propto \left(-\nabla o_{nm}\right) = \mathbf{x}_n - \mathbf{w}_m \tag{6.11}$$

In SOM, the update rule is defined by co-operating the competitive learning mechanism mentioned above

$$\Delta \mathbf{w}_m = \upsilon(\mathbf{x}_n - \mathbf{w}_m) \tag{6.12}$$

where υ is composed of two parts, one being associated with a decaying learning rate and the other being associated with the relationship between the target neuron and a neuron in the neighbourhood. The decaying learning rate is a positive real number

$$\eta^t = \eta^0 \left(1 - \frac{t}{T}\right) \tag{6.13}$$

where $\eta^0 \in (0,1)$ is the initial learning rate, η^t is the learning rate at time t and T is the maximum learning cycle. The neighbourhood relationship is defined as the distance between a neuron (not its code vector) and the target neuron. The Hamming distance or Euclidean distance can be used to quantify the relationship. The distance is converted to a rate as

$$\lambda_n^t = \exp(-\varphi(k, \phi(\mathbf{x}_n)), \forall k \in \Phi^t(\mathbf{x}_n) \tag{6.14}$$

where $\varphi(k, \phi(\mathbf{x}_n)$ is the distance between a neuron in the neighbourhood $\Phi(\mathbf{x}_n)$ and the target neuron $\phi(\mathbf{x}_n)$. The decaying neighbourhood is defined as

$$\omega^t = \min\left\{ \left[\omega^0 \left(1 - \frac{t}{T}\right) \right], 1 \right\} \tag{6.15}$$

where ω^0 is the initial neighbourhood size, which is commonly half of the size in one dimension of the two-dimensional SOM map shown in

Fig. 6.3, and ω^t is the neighbourhood size at time t. Finally, the code vector update rule is defined as below

$$\Delta \mathbf{w}_m^{t+1} = \eta^t \lambda_n^t (\mathbf{x}_n - \mathbf{w}_m^t) \qquad (6.16)$$

or

$$\mathbf{w}_m^{t+1} = \mathbf{w}_m^t + \eta^t \lambda_n^t (\mathbf{x}_n - \mathbf{w}_m^t) = (1 - \upsilon)\mathbf{w}_m^t + \upsilon \mathbf{x}_n \qquad (6.17)$$

Defining

$$\upsilon^* = \frac{(\mathbf{x} - \mathbf{w})^\mathrm{T} \mathbf{x}}{\| \mathbf{x} - \mathbf{w} \|^2} \qquad (6.18)$$

From Fig. 6.6 we can see that the update of $\mathbf{w}_m^t$ is more on the $\mathbf{w}_m^t$ side if $\upsilon < \upsilon^*$ and more on the $\mathbf{x}_n$ side if $\upsilon > \upsilon^*$. When $\upsilon = 0$, $\mathbf{w}_m^{t+1} = \mathbf{w}_m^t$ meaning no learning at all. When $\upsilon = 1$, $\mathbf{w}_m^{t+1} = \mathbf{x}_n$ meaning that code vectors always take the positions of input vectors. A learning process can never converge. From Fig. 6.6, we can see that a careful selection of the learning parameter is important to an efficient learning process. The learning rate is therefore normally smaller than 0.3.

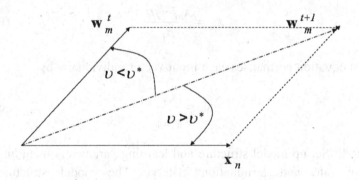

Fig. 6.6. An illustration of the impact of learning rate on the update of code vector.

The procedure described below is used to make a SOM model for a given data set.

Step 1: Pre-process data. SOM only accepts numeric input vectors. If data are non-numeric, some metric must be introduced before using SOM. Some techniques for handling non-numeric biological data will be discussed below. The next important subject in using SOM is data normalisation. There are three alternatives for data normalisation. The first is using linear scaling. The second is using normal distribution conversion. The third is self-normalisation. With linear scaling, each input vector is scaled by

$$\mathbf{x}_n = \frac{\mathbf{x}_n - col\min(\mathbf{X})}{col\max(\mathbf{X}) - col\min(\mathbf{X})} \qquad (6.19)$$

where $\mathbf{X}$ is a matrix of input vectors in which an input vector ($\mathbf{x}_n \in \Re^d$) is placed in a row, $col\min(\mathbf{X})$ is column-wise minimisation, and $col\max(\mathbf{X})$ is column-wise maximisation. Using normal distribution conversion, we need to calculate the mean and standard deviation of a variable as below

$$\mu_i = E(X_i) \qquad \sigma_i = \text{var}(X_i) \qquad (6.20)$$

where X_i means the ith variable of the input vector data. From this, we calculate

$$x_{ni} = \frac{x_{ni} - \mu_i}{\sigma_i} \qquad (6.21)$$

The last equation normalises each input vector individually by

$$\mathbf{x}_n = \frac{\mathbf{x}_n}{\|\mathbf{x}_n\|} \qquad (6.22)$$

Step 2: Set up model structure and learning parameters including the learning rate and termination criteria. The model structure is parameterised by the number of neurons (cells) and neuron layout (rectangular or hexagonal map). There are three commonly used termination criteria. The first is the maximum learning cycles, i.e. *T*. A learning process will be terminated if $t \geq T$. The second is the learning error defined as below

$$E = \frac{1}{N} \sum_{n=1}^{N} \left\| \mathbf{x}_n - \phi(\mathbf{x}_n) \right\|^2 \qquad (6.23)$$

A learning process will be stopped if $E \leq \varepsilon$ where $\varepsilon > 0$ is a small number defined by the user. The third is the model parameter stability which measures if model parameters have been in the status of saturation while the first two criteria are still not satisfied. It is defined as the distance between model parameters in two consecutive iterations

$$S = \frac{1}{M} \sum_{m=1}^{M} \left\| \mathbf{w}_m^t - \mathbf{w}_m^{t-1} \right\|^2, \forall t > 1 \; \& \; t \leq T \qquad (6.24)$$

When $S \leq \delta$ where $\delta > 0$ is a small number defined by the user, we complete a learning process. The third criterion is introduced in case T is too large and ε is too small for an application.

Step 3: Initialise code vectors. All code vectors are assigned random values.

Step 4: Update code vectors iteratively. Here equation (6.17) is used repeatedly until one of three conditions is satisfied.

When using the self-normalisation technique, we need to take care of data distribution. If data in different dimensions have large differences, it will not produce useful data for SOM. Figure 6.7 shows such a case, where the original two-dimensional data have large differences between two dimensions. The first two normalisation techniques can maintain the data structure, but the last one results in distorted distribution.

6.4 Using SOM for classification

Like other unsupervised learning approaches, one of the ultimate goals of VQ and SOM is to develop a system with classification rules. This means that a VQ or a SOM learning process assumes that the underlying data structure or topological structure is related to data classification. For instance, a SOM model is well built using well-prepared data of both disease-related and disease-free gene expression profiles. Although data are not labelled, the model will generate a map on which two classes of gene expression profiles should be well separated. Assuming that this is the case, a post-analysis of the SOM output map makes sense for exploring classification rules. The common procedure is to label each neuron or a code vector onto which some input vectors have been mapped. According to the statistical property of these input vectors, the neuron or the code vector can be used as a prototype (classification rule). If a novel input vector has the smallest distance with a code vector which has been labelled, a prediction of the biological property of the novel input vector can be made.

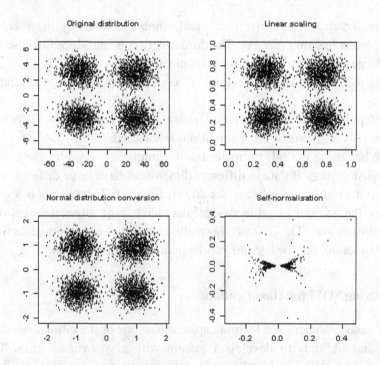

Fig. 6.7. An illustration of three normalisation techniques for data with four clusters.

6.5 Bioinformatics applications of VQ and SOM

VQ has been used in bioinformatics, but not very often so far. In analysing gene expression data, it has been found that combining fractal dimension and discrete wavelet decomposition with VQ can improve the clustering accuracy compared with the conventional clustering algorithm [199]. In analysing microarray data of human lymphoma, VQ can be extended to discriminant analysis for improving the quality of feature clustering, hence leading to meaningful signatures [200]. Meanwhile, SOM has been very intensively applied to bioinformatics. We are going to classify these applications in terms of subjects. Three subjects are discussed. They are sequence analysis, gene expression data analysis, and metabolite data analysis.

6.5.1 *Sequence analysis*

Sequence homology alignment approaches are used to find which experimentally annotated database sequences are significantly similar to a novel sequence. From this, the prediction of protein structures or functions can be made for the newly sequenced protein. It is understood that proteins with similar structures or functions should have some commonly reserved motifs within sequences. Therefore unsupervised learning approaches can be used to discover the patterns constructed by motifs. With the belief that they are normally short segments within sequences, *motifs* as words are extracted from sequences. Features are constructed based on the motif frequencies. The extracted features are then used to construct a SOM model. Based on the trained SOM model, predictions can be made for any newly sequenced protein [201]. Given a set of experimentally annotated sequences $s = \{ s_n \}, \forall n \in [1, N]$, k-mer (k ranges from 4 to 16 in Hanke et. al.'s work [201]) motifs are extracted which are denoted by $\mathcal{D} = \{ \mathbf{x}_n \}, \forall n \in [1, N]$. $\mathcal{D}$ is then a set of input vectors. Based on $\mathcal{D}$, a SOM model is constructed, $\text{SOM} : \mathcal{D} \mapsto \mathcal{M}$. $\mathcal{M}$ is a set of optimal or near optimal code vectors. The output map shown in Fig. 6.3 is then labelled based on the annotated information of sequences which have mapped to each neuron (cell) of the map. This means that $\mathcal{M} \Leftrightarrow \mathcal{F} = \left\{ \alpha_m \right\}, m \in [1, M]$, where $\mathcal{F}$ is a collection of annotated structural and functional information for each neuron (cell).

Using the same approach, the same number of features (input vector $\mathbf{x}_{N+1}$ with $N+1$ meaning a sequence beyond the collections in s) of a new sequence(s) is formed in the same way when generating $\mathcal{D}$. By inputting $\mathbf{x}$ to $\mathcal{M}$, a winner is found for it and is denoted by $\varsigma_m \in \mathcal{M}$. $\alpha_m \in \mathcal{F}$ is then the prediction for $\mathbf{s}$. A similar technique has also been used for classifying prokaryotic and eukaryotic proteins [202], and for analysing DNA sequences [203]. One challenge in biology is species diversity. Unsupervised learning approaches can be well armed for exploring the unknown diversity hidden in data. For this reason, 60,000 gene sequences of 29 bacterial species have been coded using principal component analysis and analysed using SOM leading to significant findings of the species diversity [204]. Such a type of applications is also referred to as alignment-free protein classification [152].

SOM can be used for short sequence segments (or peptides) data analysis projects. For instance, characterising functional peptides is such an application. Peptides are coded using a specific technique. The coded peptides are treated as numeric input vectors which are used to train a SOM model. Each neuron (cell) of a trained SOM model can be labelled according to the status of functional peptides which are mapped to it. A SOM learning process can be treated as a process of completing a discrete mapping from input vectors to a two-dimensional space shown in Fig. 6.3. The SOM output map is then treated as a feature map. A labelled feature map can then be used for analysing novel peptides or for prediction. This is exactly the same as what is mentioned above for analysing protein sequences.

In using the second generation of DNA sequencing technology, we will have many fragments of a DNA genome sequence. Except for the species diversity, the other challenge is the assembly of fragments of non-sequenced species. Self-organising map has been adapted to hierarchically growing hyperbolic SOM to cluster variable-length DNA fragments. From this, DNA fragments from different species are classified and visualised [205]. In constructing such a hierarchical growing hyperbolic SOM, k-mer motif frequencies are used as features. Figure 6.8 shows the diagram of the model.

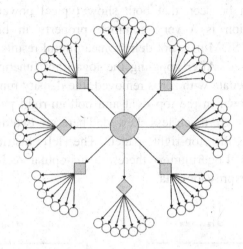

Fig. 6.8. An illustration of hierarchically growing hyperbolic SOM. There are three layers from the middle to the most outer. The neurons are expanded layer by layer [205].

6.5.2 *Gene expression data analysis*

One significant benefit of SOM in analysing gene expression data is twofold. First, like other unsupervised learning algorithms, SOM is able to reduce the dimensionality of a data set. From this, topological structure can be visualised. Compared with other unsupervised learning algorithms, SOM has the advantage of handling nonlinear data. Second, unlike other unsupervised learning algorithms, SOM is able to partition and visualise data at the same time. This is extremely welcome in analysing gene expression data which normally have very high dimensionality [203, 206, 207].

Because of the distinct property of gene expression data, the application of SOM to gene expression data requires specific techniques. First, the magnitudes of gene expression within one data set can be at different scales, i.e. from a few hundred to a few hundred thousand. Figure 6.9 shows four density functions of the *Burkholderia pseudomallei* gene expression data. The top-left panel shows the density of the original raw data where no data pre-process has been done. The bottom-left panel shows the density function of the original raw data in which zeros are removed. It can be seen that both show typical power distributions. Power distribution is a very common property in biological data. Because of this, SOM may not deliver meaningful results if data are not well pre-processed. After applying the logarithm function on the raw data and the raw data with zeros removed, the density functions of these data sets are shown in the top-right and bottom-right panels. It can be seen that the data are much less skewed. There are two distinct separate distributions in the top-right panel. The left distribution mainly represents zeros. Logarithm is therefore a popular technique used in analysing gene expression data.

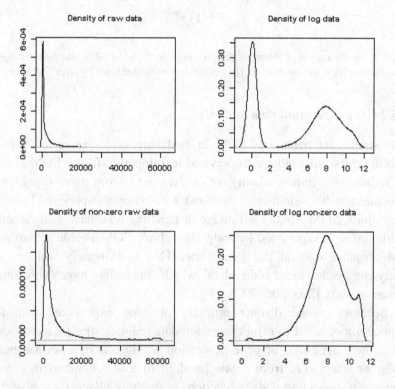

Fig. 6.9. An illustration of data skew. The data set used in this illustration is the *Burkholderia pseudomallei* gene expression data.

The second commonly used technique is differentiating data. One of the purposes of using SOM is to explore the hidden pattern in gene expression data so as to explore the intrinsic qualities of a biological system. To do this, it is necessary to determine which subset of genes are significantly positively or negatively differentiated between experimental samples and control samples. If the control group and the experimental group are well paired, using subtraction with each pair, we can form a secondary data set, i.e. differential data for analysis. Where $\mathbf{x}_n^e \in \mathfrak{R}^d$ and $\mathbf{x}_n^c \in \mathfrak{R}^d$ denote the nth gene which has been experimented on in both experimental and control groups, the differential gene is defined by

$$\mathbf{z}_n = \mathbf{x}_n^e - \mathbf{x}_n^c \text{ or } \mathbf{z}_n = \frac{\mathbf{x}_n^e - \mathbf{x}_n^c}{\mathbf{x}_n^e + \mathbf{x}_n^c} \tag{6.25}$$

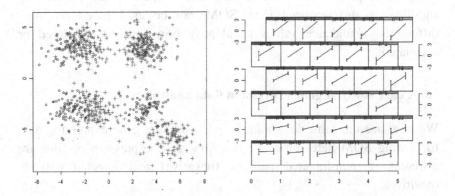

Fig. 6.10. An experiment of applying SOM to artificial gene expression data. The SOM model is composed of 30 neurons with a hexagonal array. The left panel shows the original data. The circles represent one group and the crosses represent the other group. The right panel shows the differential patterns of all 30 neurons after training a SOM model. Both control and experimental groups have four clusters of Gaussian distributions each having 100 input vectors.

Note that d is the number of samples (for instance, disease-related and disease-free patients) in the control and experimental groups. Rather than working on the raw data, a differential SOM model can be generated using $\{\mathbf{z}_n\}$. Based on differential data, a well-trained SOM model will visualise the clustering of differential genes in terms of their differentiation magnitudes. Figure 6.10 shows an experiment of applying SOM to artificial gene expression data. It can be seen that the differential patterns are well grouped into different neurons. Meanwhile neighbouring neurons show similar differential patterns.

6.5.3 *Metabolite data analysis*

Equipped by Liquid Chromatography Tandem Mass Spectrometry, metabolites which play an important role in cellular functions can be accurately identified with the resolution up to three or four decimal points. Based on this resolution, an ion can be mapped to a chemical formula, hence a compound which is stored in a database. However, over 80~90% ions may not be mapped to any compound. In order to explore more information about these ions, particularly significantly differentiated ions, SOM can be used based on ions' differential abundance values to identify how ions are clustered or correlated.

6.6 A case study of gene expression data analysis

We now use the same *Burkholderia pseudomallei* gene expression data for testing SOM here. First, we test how gene expression profiles are different in non-infected patients (negative) and infected patients (positive).

Negative and positive gene expression profiles are then separately fed into SOM leading to two SOM output maps shown in Fig. 6.11. It can be seen that the two maps show different gene expression profiles. In the positive map (the right panel in Fig. 6.11), we find most neurons show high profiles. However, in the negative map (the left panel), most neurons display low profiles.

Next, we use all data to test if SOM can discover the gene expression profile structure according to the classification of patients. A SOM model with 100 neurons is constructed and the output map is used to display the status of mapped patients. Figure 6.12 shows the result

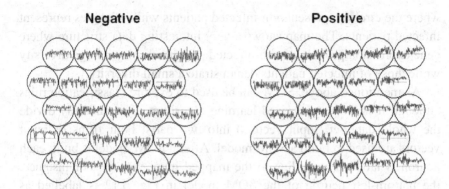

Fig. 6.11. An illustration of using SOM to analyse gene expression profiles of non-infected patients (negative) and infected patients (positive). 25 neurons are used for the test.

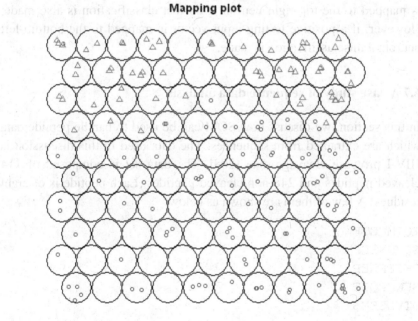

Fig. 6.12. An illustration of using SOM to study the general data structure of a gene expression data set.

where the circles represent non-infected patients while triangles represent infected patients. The map shows a very interesting data structure where gene expression profiles of non-infected patients display a large diversity while those of infected patients demonstrate a small diversity.

As mentioned above, SOM can be used to explore classification rules although it is an unsupervised learning algorithm. We randomly divide the whole data set (input vectors) into two parts. Four fifths of input vectors are used to build a SOM model. After a SOM model is built, each neuron is labelled according to the mapped input vectors. For instance, the bottom-left neuron of the SOM model in Fig. 6.12 is labelled as negative while the top-right neuron of it is labelled as positive. We then feed the remaining one fifth of input vectors (referred to as testing input vectors) into the SOM model to see to which neurons they will be mapped. If a negative testing input vector is mapped to the bottom-left neuron, a correct classification is made. If a positive testing input vector is mapped to the top-right neuron, a correct classification is also made. However, if a positive testing input vector is mapped to the bottom-left neuron, a misclassification is made.

6.7 A case study of sequence data analysis

In this section, we discuss how SOM can be used to handle peptide data which are extracted from sequences. The data used in this discussion is HIV-I protease cleavage data [208]. The data set is composed of 114 cleaved peptides and 248 non-cleaved peptides. Each peptide is of eight residues. A few of them are shown as below:

```
TQIMFETF
GQVNYEEF
PFIFEEEP
SFNFPQIT
DTVLEEMS
```

Each of eight residues in a peptide is one of 20 amino acids. They are therefore non-numeric. In order to make them usable to SOM, a coding process is needed. The orthogonal sparse coding [209] is one of the most used. Based on this coding technique, each amino acid is coded by a 20-bit long binary string. In this string, one bit is assigned a value one leaving all other bits as zeros. For instance, an Alanine is coded by 0000000000, 0000000001 and a Cystine is coded by 0000000000, 0000000010, etc. The coded data can then be fed into a SOM model for both data visualisation and knowledge discovery. Figure 6.13 shows the output map of a SOM model with 100 neurons for the data. Each neuron has an associated code vector which can be decoded. The decoded code vector can be treated as a motif (mean vector) for all the peptides (input vectors) mapped onto it. Based on the fraction of cleaved peptides over all peptides mapped onto a neuron, a contour is formed to visualise which motifs are contributing to HIV-1 cleavage and which are not.

Fig. 6.13. An illustration of a SOM model with 100 neurons for the HIV data.

Figure 6.14 shows how SOM is used to visualise the way in which cleaved and non-cleaved peptides are distributed. It can be seen that most neurons generally have a single class of peptides, either being non-cleaved peptides or cleaved peptides. This demonstrates that the internal topological structure hidden in peptides can be explored to form significant classification rules. The rules can be well used in novel cleavage site prediction and drug design.

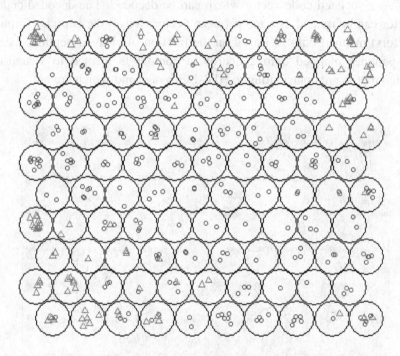

Fig. 6.14. An illustration of how the distribution of peptides can be visualised using SOM.

Summary

This chapter has focused on Self-organising map, a neural learning algorithm developed on the basis of vector quantisation. The principles, the structure, and the learning algorithm of SOM have been discussed. Because most biology experiments are in nature exploring unknowns, the basic classification rules are normally lacking. Pattern analysis and topological structure re-construction are then a particularly important subject in bioinformatics. Having understood that biological data sets are getting larger and larger, a challenge for SOM is how to handle data efficiently. For instance, training a SOM model with 10,000 neurons with 40,000 input vectors of 20 dimensions may take a couple of weeks to complete. One issue is to improve the learning algorithm and the other issue is to combine knowledge of machine learning and biology to efficiently pre-process the data before using SOM. For instance, genes which are completely silent may not make any difference in analysing how genes are differentiated in a sample. Pre-filtering the data may save much computing time.

Chapter 7

Introduction to Supervised Learning

Supervised learning is a subject in which a model is estimated from data for mapping explanatory variables to predictive variables. The explanatory variables in bioinformatics often refer to genotypic data which are used to describe underlying properties of a set of molecules within an organism. The predictive variables in bioinformatics are often used to describe phenotypic data which are observed. One of the most important targets in scientific research is to explain phenotypic phenomenon using genotypic data. This is very similar to most of our human intelligence activities, i.e. finding reasons that explain observed events, for instance, interpreting what causes the climate to change, understanding the genetic reasons leading to aging disease, deducing the underlying regulation of the financial market, and studying species diversity. Supervised learning is then searching for the most appropriate explanatory variables for interpreting predictive variables using a model. Such a model is often called a predictive model. Whenever such a model is established, its predictive function can be used for inference. This chapter focuses on the general concept of supervised learning. The discussion covers general concepts, rules, data organisation, model evaluation, and model feasibility.

7.1 General concepts

Various unsupervised learning approaches and algorithms have been discussed in the previous chapters. The main objective of unsupervised learning is to discover unknown knowledge of a data structure. The use of unsupervised learning is based on the nature in many scientific research projects that there is no definite classification law available and we are required to find it. For instance, given a set of genome sequences from different species, how can we determine the diversity among them?

A supervised learning process represents a different concept where we are interested in mapping one data point to the other data point, specifically finding how explanatory variables can interpret predictive variables. Here explanatory variables refer to the data which describe the causes for interpreting observed phenomenon. Predictive variables are directly associated with observed phenomenon. Having understood that observed phenomenon often contain variations caused by many man-made and natural causes, predictive variables are treated as the unbiased phenomenon we are searching for. In bioinformatics, explanatory variables often refer to genotypic data while predictive variables refer to observed phenotypic data. The major task of bioscientists is to find causes to interpret the observed phenotypic data. For instance, when we find the diverse reactions to a drug among patients we may need to investigate the genotypic reasons that can explain the diversity. When we find diverse growth rates in plant mutants we may need to investigate which metabolites and which metabolite pathways contribute to the diversity. In science this is an induction process, i.e. from data to knowledge. In machine learning it is called a learning process or a supervised learning process.

Note that this mapping is not the final goal of a supervised learning process. Ultimately we are interested in how classification rules can be established for the prediction of unknowns. This is a deduction process, i.e. from knowledge to data (new data). The prediction of biological diversity [204, 210, 211], cellular function [212-215], protein modifications [216-238], and the diagnosis of cancers using microarray technology [239-242] are examples of this.

In predicting species diversity, some data are composed of genome sequences (part or whole) and other data are diversity measurements. The genome sequence data are regarded as genotypic data while the diversity is treated as phenotypic data. A computer model built using a machine learning approach can be used to predict diversity whenever new genome sequences are obtained. In predicting cellular functions, the genotypic data can be gene expression profiles or genomic sequences while the phenotypic data are the cellular functions including cellular localisation, protein-protein interaction, and signal transduction. In predicting protein modifications, the input data are the sequence segments which normally have less then 20 residues and the output data are the modification statuses, i.e. having a phosphorylation site or not. When we build a model for cancer diagnosis using microarray data the genotypic data are genes with differential profiles from normal and abnormal tissues. The phenotypic data or model outputs are the cancer statuses, i.e. having cancer or not.

All aim to construct a mapping (predictive) function from explanatory variables to predictive variables. The mapping function is normally unknown. In some cases, function parameters are unknown and in other cases both function parameters and functional structure are unknown.

7.2 General definition

A data set which is used for a supervised learning process is denoted by $\mathcal{D} = \left\{ \mathbf{x}_n, t_n \right\}_{n=1}^{N}$, where $\mathbf{x}_n \in \Re^d$ is called an input vector representing explanatory variables and $t_n \in \Theta$ (Θ is a set of phenotypic status) is called a target variable for a predictive variable. The number of predictive variables can be easily extended to multiple ones. It is assumed that $\mathcal{D}$ is randomly sampled from a space ($\Re^d \times \Theta$) satisfying an unknown function

$$\mathcal{D} : f(\mathbf{x}) \mapsto t \tag{7.1}$$

Note that the number of data points in $\Re^d \times \Theta$ is in general infinite and the number of data points satisfying $f(\mathbf{x})$ is infinite while the size of $\mathcal{D}$ is finite, $\left| \mathcal{D} \right| << \infty$. What this means is that there are infinite data points of a data space spanned by a function. However, we can only collect a finite number of data points used as input vectors. The job required is to estimate $f(\mathbf{x})$ (the estimated version of $f(\mathbf{x})$ is denoted by $\tilde{f}(\mathbf{x}_n)$) using $\mathcal{D}$

$$\tilde{f}(\mathbf{x}_n) \mapsto t_n, \forall \left\{ \mathbf{x}_n, t_n \right\} \in \mathcal{D} \tag{7.2}$$

Because the size of $\mathcal{D}$ is limited the true function $f(\mathbf{x})$ may not be exactly estimated. We then have to define a criterion by which we can find the best estimated function for $f(\mathbf{x})$

$$\tilde{f}(\mathbf{x}) = \arg\min \left| g(\mathbf{x}) - f(\mathbf{x}) \right|, \forall\, g(\mathbf{x}) \tag{7.3}$$

where $g(\mathbf{x})$ represents a possible candidate function.

Before discussing detailed supervised learning algorithms, we need to classify supervised learning algorithms based on two important criteria.

The first depends on the status of t. If it is continuous ($t \in \mathfrak{R}$), we are dealing with a regression analysis problem. For instance, a regression model can be used to link gene expression polymorphisms of interleukins to oral squamous cell carcinoma [243], to link metabolism to growth-related properties based on metabolite profiles [244], or to infer gene expression dynamics [245]. If the target variable t is discrete, i.e. $t \in I$ or $t_n \in \{\text{complete inhibitive, low inhibitive, no inhibition}\}$ we then deal with a classification analysis problem. Many bioinformatics subjects fall into this classification analysis category. They include diversity prediction, cellular function prediction, protein modification prediction and cancer diagnosis using gene expression profiles.

The next issue is to do with linear supervised models versus nonlinear supervised models. A linear model has the benefit of intuitiveness, i.e. a model is easy to interpret. For instance, a linear model relating cancer status and two gene expression profiles can be defined as

$$y = w_0 + w_1 x_1 + w_2 x_2 \tag{7.4}$$

where $\{x_1, x_2\} \in \mathfrak{R} \times \mathfrak{R}$ are the expression profiles of two genes, $y \in \{\text{no, yes}\}$ indicates whether a patient has cancer or not and $\{w_0, w_1, w_2\} \in \mathfrak{R} \times \mathfrak{R} \times \mathfrak{R}$ represent the model parameters. Here y is used to indicate how likely a patient has cancer. If a model has been built and $w_1 \gg w_2$, the first gene then has higher contribution to cancer development compared with the second gene. However, all linear models have a fatal limitation that they are unable to handle nonlinear data. Nonlinear mapping functions are then sought for analysing nonlinear data. As mentioned above that in most bioinformatics projects no definite classification law is available before a study starts, a challenging subject is to determine a nonlinear model function which is generally difficult to estimate. An alternative is to construct a mapping function working as a "black-box". In a black-box mapping function, interpreting which genes are more important than others may not be easy. For instance, in a mapping function

$$y = f(x_1, x_2, \mathbf{w}) \qquad (7.5)$$

The major objective is to find how inputs (x_2 and x_2) are accyurately mapped to the outputs (y). Note that $\mathbf{w}$ is a set of model parameters. Most machine learning algorithms are basically nonlinear.

7.3 Model evaluation

As mentioned above, it is difficult to find the true model in practice. Among many candidate models, model evaluation is then an important issue related to model selection, i.e. evaluating which model is the best. Figure 7.1 shows an example.

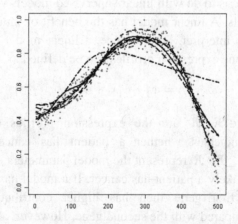

Fig. 7.1. An illustration of fitting a data set using different models which vary in terms of how well they fit the data.

The univariate function is defined as

$$f(x) = 0.2x + 0.25x^2 - 0.03x^3 + \mathcal{N}(0,0.3) \qquad (7.6)$$

where $\mathcal{N}(0,0.3)$ is a noise added with a zero mean and a standard deviation of 0.3. 500 data points are randomly generated. For this limited data size, different models fit the data differently, i.e. some fitting the data better than others. In order to evaluate these models, a quantitative criterion must be used.

Because a supervised learning model can be a regression one or a classification one, the evaluation strategy then varies. For most regression models, an error function can be defined as the mean-square error as below

$$\frac{1}{N}\sum_{n=1}^{N}(t_n - y_n)^2 = \frac{1}{N}\sum_{n=1}^{N}(t_n - f(\mathbf{x}_n))^2 \qquad (7.7)$$

In order to make models of different data comparable, the normalised mean-square error can be used. It is defined as below

$$\frac{\sum_{n=1}^{N}(t_n - y_n)^2}{\sigma_t^2} \qquad (7.8)$$

where σ_t^2 is the data variance. Correlation between the targets and model outputs can also be used. It is defined as

$$\rho(T,Y) = \frac{\sum_{n=1}^{N}(t_n - \mu_t)(y_n - \mu_y)}{\sigma_t \sigma_y} \qquad (7.9)$$

where σ_y^2 is the variance of model output, μ_t and μ_y are the mean values of targets and model outputs.

For classification, there are two approaches, one is called single point estimation and the other is called robustness estimation or probability analysis of data separation. Using the single point estimation, the accuracy of each class is estimated by using one pre-defined threshold for classification. For instance, in a two-class classification problem we may make decisions as below

$$d(y) = \begin{cases} 0 & if \ y < T \\ 1 & otherwise \end{cases} \tag{7.10}$$

where T is the threshold, y is the model prediction and $d(y)$ is the decision made based on the comparison of y against the threshold. The estimation is completed using a confusion matrix in which equation (7.10) is repeatedly used for all data points and a summarisation is made. For instance a model for a two-class classification problem is able to accurately predict 90 of 100 positive input vectors and 80 of 100 negative input vectors. A confusion matrix is formed as below

	Negative	Positive	Percent
Negative	80	20	80%
Positive	10	90	90%
	89%	82%	85%

The accurately predicted positive and misclassified positive are called true positive (TP) and false negative (FN). The accurately predicted negative and misclassified negative are called true negative (TN) and false positive (FP). In the above table, TN and FP are 80 and 20, respectively. TP and FN are 90 and 10, respectively. The prediction accuracy of the negative class is known as the specificity and the prediction accuracy of the positive class is known as the sensitivity. The specificity is defined as

$$Spe = \frac{TN}{TN + FP} \tag{7.11}$$

and the sensitivity is defined as

$$Sen = \frac{TP}{TP + FN} \tag{7.12}$$

The total prediction accuracy is defined as

$$Tot = \frac{TN + TP}{TN + FP + TP + FN} \qquad (7.13)$$

Prediction powers [246-249] can be used to evaluate the confidence of trusting a prediction. The prediction powers are different from prediction accuracies. The negative predictive power (NPP) measures how likely a negative prediction is true. It is calculated by the fraction of correctly identified negative input vectors over the total predicted negative input vectors. The positive predictive power (PPP) then measures the probability that a positive prediction is true. This is calculated by the fraction of correctly identified true positive input vectors over the total predicted positive input vectors. They are defined as

$$NPP = \frac{TN}{TN + FN} \qquad PPP = \frac{TP}{TP + FP} \qquad (7.14)$$

The receiver operating characteristics (ROC) analysis [250] can only be used in a two-class classification problem for system robustness evaluation. It is used to analyse how likely the predictions of two classes are to be well-separated. In particular, the areas under ROC curves (AUR) are normally used as a quantitative indicator of model robustness. Because ROC is used for analysing the separation quality between predictions of two classes, AUR is also named as the probability of separation. When conducting ROC analysis, we need to vary the threshold (see equation (7.10)) by which classification is made. Because of the change of the threshold, the sensitivity and specificity change. For each threshold there is a pair of values for specificity and sensitivity. For many thresholds, there are many pairs of them. We then map these two points into a two-dimensional space using the false alarm rates (1 - specificity) as the horizontal axis and the sensitivity as the vertical axis. The points are connected to form an ROC curve. Figure 7.2 shows the ROC analysis for two data sets of one dimension. One shows a reasonable separation between two classes of data and the other shows a larger overlap between two classes of data. It can be seen that the ROC of data set 1 is closer to the top-left corner compared with that of data set 2. This shows that data set 1 is better than data set 2 in terms of separation capability between two classes of data. The AUR of data set 1 is then larger than that of data set 2.

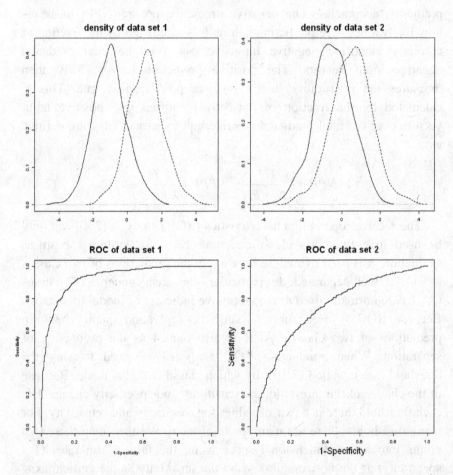

Fig. 7.2. An illustration of ROC analysis. The solid lines and the dashed lines in the top panels represent two classes of data. The bottom panels show the ROC curves for them. The R ROCR package [251] is used to build ROC curves. In the ROC curves, the horizontal axes represent 1 – specificity and the vertical axes represent sensitivity.

7.4 Data organisation

In order to deliver a model with a proper and unbiased estimation of prediction accuracy, proper data organisation must be considered before starting computer simulation and model construction using any machine learning algorithm.

Because data collection is a random process the data collected one time will not be identical to data collected the next time. Model construction must consider such a variation. As mentioned above, the observed phenotypic data are likely noise-contaminated data. There is then a critical learning problem. How can we distinguish between true data and noise data?

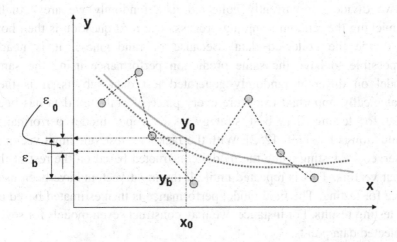

Fig. 7.3. A diagram showing the danger when a function (model) fits data too well. The input variable is denoted by x, the output variable is denoted by y. A new data point is denoted by x_0. The dots are the collected data for estimating a function. The true function by which data are collected (randomly sampled) is denoted by a thick smooth line. The estimated too-good function is denoted by the dashed line which connects each collected data point, perfectly fitting the data. The estimated not-too-bad function is denoted by a thick dotted line.

If we treat all the collected data as true data a model may fit the collected data very well but has little capability to generalise. Figure 7.3 shows such a situation where a function (model) is built to fit the collected data (x, y) pairs very well without any error. When we have a new datum denoted by x_0, a large prediction error ε_b occurs when using the too-good function whose prediction is y_b. The prediction is y_0 and the prediction error is smaller (ε_0) using a not-too-bad function. However, the not-too-bad function has a large error compared with the too-good function, in fitting the collected data. The prediction capability for new data is often referred to as the generalisation capability. A model which has less prediction error for the new data (the not-too-bad function illustrated by the dotted line in Fig. 7.3) is called a model with a good generalisation capability.

However, how can we estimate a model's generalisation capability before seeing new data? A common methodology we adopt is to use the current collected data. We know that data collection is a random process. If we divide our currently collected data randomly we are actually mimicking the random sampling process. The next question is then how to divide the collected data. Because of randomness, it is nearly impossible to have the same prediction performance using the same model on different randomly generated test data subsets. It is then strategically important to ensure every piece of collected data has been used for testing. The best strategy of a proper model performance evaluation is Jackknife [252]. With the Jackknife test, one input vector is reserved for testing using the model constructed based on the rest of the input vectors. This is repeated until all the input vectors have been used once for testing. The final model performance is then estimated based on all testing results. For instance, we may construct seven models for seven collected data points.

When data size gets very large, Jackknife becomes computationally infeasible. In this situation, k-fold cross-validation is used. Data are randomly divided into k-folds. K models are constructed. Each of them uses one fold of input vectors as test data while the rest are used for model construction. The final model performance is estimated based on these k sets of testing results.

However, it must be noted that even using k-fold cross-validation two separate runs may still generate different performance estimations. If computational time is feasible, N k-fold cross-validations can be carried out and the final performance is a mean of each of these N runs.

7.5 Bayes rule for classification

A practical issue related to classification analysis is the confidence of a prediction. If a model provides probabilities of different classes, the Bayes rule can be used for decision-making. Suppose there are K classes, the probability (or conditional probability) that an input vector belongs to the kth class is denoted by $p(\mathbf{x} \mid k)$ and the *a prior* probability of the kth class is $p(k)$. The post-probability (given K classes) is defined as

$$p(m \mid \mathbf{x}) = \frac{p(\mathbf{x} \mid m) p(\mathbf{x} \mid m)}{\sum_{k=1}^{K} p(\mathbf{x} \mid k) p(\mathbf{x} \mid k)}, \forall m \in [1, K] \qquad (7.15)$$

A decision is made through maximising the post-probability denoted by $p(m \mid \mathbf{x})$.

Summary

In this chapter, general concepts and rules of supervised learning have been discussed. Model evaluation and data organisation strategies have also been discussed. All these are fundamental to the following chapters where several commonly used machine learning algorithms for bioinformatics will be discussed. It must be noted here that these practices are important for generating unbiased, accurate and precise models and should not be ignored in experimental design.

Chapter 8

Linear/Quadratic Discriminant Analysis and K-Nearest Neighbour

In this chapter, two important machine learning approaches which can be used for supervised learning are introduced. One is called discriminant analysis including linear discriminant analysis and quadratic discriminant analysis and the other is called the K-nearest neighbour algorithm. Linear or quadratic discriminant analysis is a simple learning algorithm which has the advantages of simplicity and intuitiveness. The K-nearest neighbour algorithm has the advantage of low learning cost. This chapter will discuss the principles of these algorithms and the procedures of their applications to bioinformatics.

8.1 Linear discriminant analysis

Linear discriminant analysis, also referred to as Fisher discriminant analysis (FDA), is a simple algorithm which has been widely used in many areas [158, 160, 253]. FDA aims to find a linear function which linearly combines independent variables using a set of weights (model parameters) to determine the property of a dependent variable. The linear function is also called a hyper-plane which separates two classes of input vectors. The hyper-plane is called the decision boundary or surface while the linear function is called a linear classifier.

The basic requirement of FDA is that we assume that we know the function form in advance for a data set. The linear function used in FDA is defined as

$$f(\mathbf{x}) = w_0 + \mathbf{w} \cdot \mathbf{x} = w_0 + \sum_{i=1}^{d} w_i x_i \qquad (8.1)$$

where $\mathbf{x} \in \Re^d$ is an input vector, $\mathbf{w} \in \Re^d$ is a weight vector (model parameters), and w_0 is a bias.

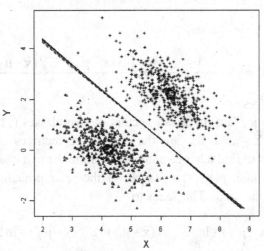

Fisher discriminant analysis

Fig. 8.1. An illustration of FDA for a data set of two-dimensional input vectors. The horizontal axis and the vertical axis represent the two dimensions, i.e. X and Y. The triangles and crosses represent two classes of input vectors. For this data set, ten FDA models are built by randomly sampling four fifths of the data. The lines which separate these two data swarms are the hyper-planes generated by FDA. The large circles represent the centres of two classes of input vectors.

A simple explanation of FDA is to find a hyper-plane on which $f(\mathbf{x}) \equiv 0$. This hyper-plane is also called the decision boundary, i.e. $\mathbf{x}$ belongs to one class if $f(\mathbf{x}) > 0$ while it belongs to the other class if $f(\mathbf{x}) < 0$. All the points on the hyper-plane satisfy $f(\mathbf{x}) = 0$ [158, 160, 253]. Figure 8.1 shows a simple example of FDA for two classes of input vectors in two-dimensional space.

There are two different approaches for estimating FDA model parameters. We can assume that data follow two Gaussian distributions for two classes of input vectors. The density functions of two classes are defined as

$$p(\mathbf{x} \mid \mathcal{A}) = \frac{1}{(2\pi)^{d/2} \sqrt{|\Sigma_{\mathcal{A}}|}} \exp\left(-\frac{(\mathbf{x} - \mathbf{\mu}_{\mathcal{A}})^{\mathrm{T}} \Sigma_{\mathcal{A}}^{-1} (\mathbf{x} - \mathbf{\mu}_{\mathcal{A}})}{2}\right) \qquad (8.2)$$

and

$$p(\mathbf{x} \mid \mathcal{B}) = \frac{1}{(2\pi)^{d/2} \sqrt{|\Sigma_{\mathcal{B}}|}} \exp\left(-\frac{(\mathbf{x} - \mathbf{\mu}_{\mathcal{B}})^{\mathrm{T}} \Sigma_{\mathcal{B}}^{-1} (\mathbf{x} - \mathbf{\mu}_{\mathcal{B}})}{2}\right) \qquad (8.3)$$

where $\mathbf{\mu}_{\mathcal{A}}$ and $\mathbf{\mu}_{\mathcal{B}}$ are the mean vectors of two classes ($\mathcal{A}$ and $\mathcal{B}$) of input vectors. $\Sigma_{\mathcal{A}}$ and $\Sigma_{\mathcal{B}}$ are the covariance matrices of these two classes. In order to find a hyper-plane to separate the two classes of input vectors, we assume $p(\mathbf{x} \mid \mathcal{A}) \equiv p(\mathbf{x} \mid \mathcal{B})$ if the prior probabilities of the two classes are the same. This leads to

$$(\mathbf{x} - \mathbf{\mu}_{\mathcal{A}})^{\mathrm{T}} \Sigma_{\mathcal{A}}^{-1} (\mathbf{x} - \mathbf{\mu}_{\mathcal{A}}) + \ln|\Sigma_{\mathcal{A}}| = (\mathbf{x} - \mathbf{\mu}_{\mathcal{B}})^{\mathrm{T}} \Sigma_{\mathcal{B}}^{-1} (\mathbf{x} - \mathbf{\mu}_{\mathcal{B}}) + \ln|\Sigma_{\mathcal{B}}| \quad (8.4)$$

In FDA, it is assumed that $\Sigma_{\mathcal{A}} \equiv \Sigma_{\mathcal{B}} \equiv \Sigma$. This means that both classes of input vectors have the same spreading volume. Solving equation (8.4) under this assumption leads to the hyper-plane defined as

$$\mathbf{w} = \Sigma^{-1}(\mathbf{\mu}_{\mathcal{B}} - \mathbf{\mu}_{\mathcal{A}}) \qquad (8.5)$$

In Fisher's original work, it is assumed that the hyper-plane made by weighting independent variables ($\mathbf{w} \cdot \mathbf{x}$) is able to separate two classes of input vectors if the ratio of between-class diversity over the within-class diversity can be maximised. The between-class diversity can be regarded as signal in data while within-class diversity can be treated as noise in data. This ratio is defined as

$$S = \frac{\sigma^2_{between}}{\sigma^2_{within}} = \frac{(\mathbf{w} \cdot \boldsymbol{\mu}_{\mathcal{B}} - \mathbf{w} \cdot \boldsymbol{\mu}_{\mathcal{A}})^2}{\mathbf{w}^T \Sigma_{\mathcal{B}} \mathbf{w} + \mathbf{w}^T \Sigma_{\mathcal{A}} \mathbf{w}} \tag{8.6}$$

Maximising the above equation leads to

$$\mathbf{w} = (\Sigma_{\mathcal{B}} + \Sigma_{\mathcal{A}})^{-1}(\boldsymbol{\mu}_{\mathcal{B}} - \boldsymbol{\mu}_{\mathcal{A}}) \tag{8.7}$$

It can be seen that both approach the same result. FDA can be extended to multiple classes where multiple hyper-planes will be formed [158, 160, 253].

In bioinformatics, FDA has been widely used. For instance, FDA has been used for predicting DNA methylation sites [254], for predicting phosphopeptides [255], for brain tumour diagnosis based on metabolite data analysis [256], and for identifying protein coding regions [257].

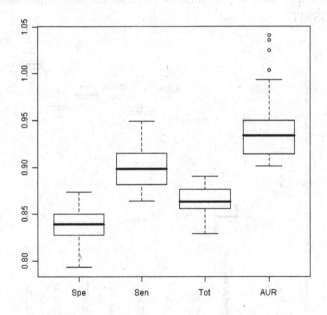

Fig. 8.2. Box plot of the prediction performances (specificity, sensitivity, total accuracy, and AUR) from 100 randomised data sets.

Here we study how FDA can be used for analysing *Burkholderia pseudomallei* gene expression data. Data size is reduced using the random forest algorithm which is discussed in the next chapter, i.e. among 214 genes, only top ten genes selected by the random forest algorithm are used for modelling using FDA. The data are randomised 100 times. Each time, five-fold cross-validation is used. The model performances (specificity, sensitivity, total prediction accuracy, and AUR) are shown in Fig. 8.2. The specificity is 86%, the sensitivity is 86%, the total prediction accuracy is 86% and AUR is 0.92.

The weights (**w**) for the top ten genes from 100 models are shown in Fig. 8.3. In these 100 random models, the gene BPSL3398 has the highest positive weight, but the gene BPSS9477 has the largest negative weight. Because the infected patients are labelled using value one and the non-infected patients are labelled using value zero, infected patients will have high expression of the gene BPSL3398 but low expression of the gene BPSS9477.

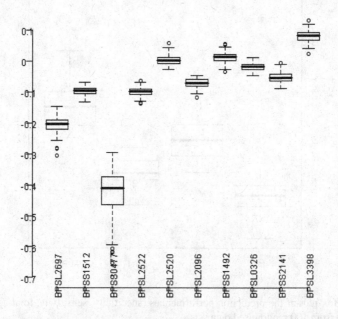

Fig. 8.3. Box plot of the weights of ten genes. The horizontal axis represents ten genes while the vertical axis represents the weights.

8.2 Generalised discriminant function

FDA is a linear machine learning algorithm. The linearity implies two learning issues. First, it is assumed that the data for modelling is linearly separable such as the case shown in Fig. 8.1. This linearity does not mean data are completely separable. Data are often not separable even when generated from two linearly separable sources. Rather they are not separable because of large overlap. For the data shown in the left panel of Fig. 8.4 FDA is unable to separate two classes of input vectors successfully because of the noise in the data. FDA is also unable to find a suitable decision surface for nonlinear data. On the right panel of Fig. 8.4, data are generated from two classes, one being points below the function y^2 and the other being points above the function. It can be seen from the right panel of Fig. 8.4 that the decision boundary made by FDA fails to separate two classes of input vectors.

Generalised discriminant analysis is also called quadratic discriminant analysis (QDA). In QDA, rather than explore the linear variables described in equation (8.1), we aim to explore non-linear variables. Here a linear variable is quantified by one independent variable while a nonlinear variable is quantified by a product of two or more independent variables. Based on this, we then have the form of QDA defined as below [158, 160]

$$y = w_0 + \mathbf{w}^T\mathbf{x} + \mathbf{x}^T\mathbf{A}\mathbf{x} = w_0 + \sum_{i=1}^{d} w_i x_i + \sum_{i=1}^{d}\sum_{j=1}^{d} a_{ij} x_i x_j \qquad (8.8)$$

The model can be re-written as below

$$y = w_0 + \tilde{\mathbf{w}}^T\mathbf{z} \qquad (8.9)$$

where $\mathbf{z} = \{ x_i \}\bigcup\{ x_{ij} \}$ and $\tilde{\mathbf{w}} = \{w_i\}\bigcup\{ a_{ij} \}$. This is in the same format of FDA described in equation (8.1). The same procedure of FDA can be applied to QDA to estimate model parameters.

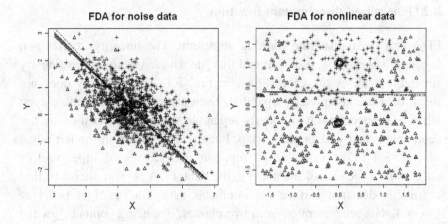

Fig. 8.4. Two examples where FDA is unable to find a suitable decision boundary. The left panel shows a case where data are generated from two linearly separable sources but highly overlapped. The right panel shows a case where data are generated from two nonlinear separable sources without overlap. The large circles represent the centres.

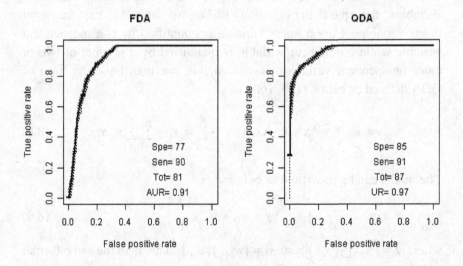

Fig. 8.5. A comparison of FDA and QDA applied to the data shown in the right panel of Fig. 8.5. The left panel shows the ROC curves of the FDA models and the right panel shows the ROC curves of the QDA models. The horizontal axes represent 1 – specificity and the vertical axes represent sensitivity.

Figure 8.5 shows a comparison of cross-validation FDA and QDA models for the data shown in the right panel of Fig. 8.5. The specificity, sensitivity, total prediction accuracy and AUR are printed in the ROC curves in Fig. 8.5 where it can be seen that QDA much outperforms FDA because of the introduction of the nonlinear variables.

In applying QDA to the *Burkholderia pseudomallei* gene expression data, we find that the performance is similar to that demonstrated when applying FDA. Although QDA introduces nonlinear variables, the capability of handling nonlinear data is still limited. This is because only the positive correlation between variables is considered, i.e. $x_i x_j$ describing positive correlation between x_i and x_j. If the classification between two classes of data depends on the negative correlation between x_i and x_j, we add more noise rather than more information.

QDA has also been used in several bioinformatics projects. For instance, it has been used for predicting protein coding regions [258], for predicting splice sites [259] and for predicting antimicrobial peptides [260].

8.3 K-nearest neighbour

K-nearest neighbour (KNN) [158] has been known as a fast learner because there is nearly no learning process at all. The principle of KNN is simple with a theoretical background. Imagine that there are K training input vectors around a query input vector within a specified volume shown in Fig. 8.6. In the Figure, the query input vector denoted by the triangle is surrounded by two classes of training input vectors. Here we use training input vectors to mean that they have already been classified. We now need to label the query input vector. An intuitive approach is to count the number of open circles and the number of filled circles. If the number of open circles is larger than that of filled circles, we can label the query input vector by the class of open circles. However, what is the theoretical background of this simple and intuitive approach?

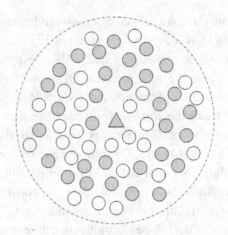

Fig. 8.6. An illustration of KNN. The open circles and the filled circles represent two classes of data while the triangle represents a query data point. The dashed circle indicates the volume centred by the query data point.

We first denote an input vector by **x** and the number of training input vectors in one class as N_k, where $k \in \{1, 2\}$ representing the open and filled circles in Fig. 8.6 respectively. We then use two simple probabilities to quantify how likely one class of input vectors is to occur in the volume which is denoted by the dashed open circle in Fig. 8.6. The simple probability estimation is defined as below

$$p(\mathbf{x} \mid k) = \frac{N_k}{N} \tag{8.10}$$

where $N = N_1 + N_2$. The posterior probability is calculated by the following equation

$$p(k \mid \mathbf{x}) = \frac{p(\mathbf{x} \mid k) p(k)}{p(\mathbf{x} \mid 1) p(1) + p(\mathbf{x} \mid 2) p(2)} \tag{8.11}$$

where $p(k)$ is the a prior probability of the kth class. If $p(1) = p(2)$ we have

$$p(k \mid \mathbf{x}) = \frac{N_k}{N} \tag{8.12}$$

If $N_1 > N_2$ the query vector is labelled as class 1. If the prior knowledge is updated to $p(1) > p(2)$ we then have

$$p(k \mid \mathbf{x}) = \frac{p(k)N_k}{p(1)N_1 + p(2)N_2} \qquad (8.13)$$

There is also another analysis that leads to the same result. Given a volume V, the probability of the kth class within it is defined as

$$p(\mathbf{x} \mid k) = \frac{N_k}{V} \qquad (8.14)$$

The posterior probability of the kth class is defined as

$$p(k \mid \mathbf{x}) = \frac{\dfrac{N_k}{V} p(k)}{\dfrac{N_1}{V} p(1) + \dfrac{N_2}{V} p(2)} \qquad (8.15)$$

It can be seen that the Bayes rule is the basis for deriving a K-nearest neighbour classification system.

Because of its simplicity, KNN has been applied to many bioinformatics projects. In the sequence domain, KNN has been used to predict transmembrane beta-barrel proteins [261] and for food protein allergenicity prediction [262]. In analysing gene expression data, KNN has been used for cancer diagnosis [263-265], toxicity analysis [266], and for identifying pathogens [267].

Like other machine learning algorithms, KNN also has a problem of model selection. For instance, Fig. 8.7 shows this dilemma. When using the inner dashed open circle as the volume in which we seek K nearest neighbours, we have found the triangle should be labelled as the filled circles. When we use the middle dashed open circle, we find that the triangle is labelled as the class of the open circles. If the outer dashed open circle is used as the volume in which to search for K nearest neighbours, the triangle is labelled as the class of the filled circles. There is therefore certainly a model selection process. For a specific model, an appropriate K number must be carefully selected.

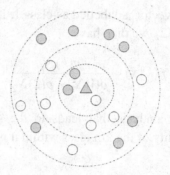

Fig. 8.7. An illustration of selecting right number of K when using KNN for a specific data set. Open and filled circles represent two classes of input vectors while the triangle represents a query input vector. Three dashed open circles show the volumes in which K nearest neighbours are sought.

The second issue is related to the distance used when searching for K nearest neighbours. Using the Euclidean distance or Manhattan distance may lead to different classification outcomes. Using the Euclidean distance from the query input vector indicated by the large open circle, the triangle is labelled as the class of filled circles. When using the Manhattan distance, the triangle is labelled as a different class. The difference can be seen in Fig. 8.8.

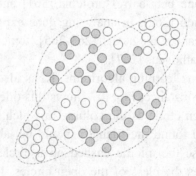

Fig. 8.8. An illustration of using different distance calculation methods leading to different labelling processes for a query data point. Open and filled circles represent two classes of input vectors while the triangle represents a query input vector. The dashed open circle and the dashed ellipse show the volumes that apply when using the Euclidean distance and Manhattan distance respectively for searching for K nearest neighbours.

The third issue related to KNN is its complexity in use. If a training data base is large, both space complexity and time complexity will be a huge burden. For a database with N training input vectors, using K nearest neighbours, KNN will need to use $N(N-1)\cdots(N-K+1)$ calculations to find the nearest neighbours.

A typical application of KNN in bioinformatics is various multiple sequence homology alignment tools such as FASTA [22] and BLAST [10]. After inputting a query sequence, the tool returns a number of database (training) sequences with ranked similarity measurements. For instance, searching BLAST for a protein P0C0R2.1 (HTH-type transcriptional regulator sarS) leads to a table shown in Table 8.1 in which the similar database sequences are listed in the first column, the significant alignment bits data are listed in the second column and the significance of similarities (e-values) are listed in the last column. The sequences are ordered from most similar ones to most dissimilar ones.

Table 8.1. The result of searching BLAST for a protein P0C0R2.1. This is a reduced table where "producing" information is removed for simiplicity.

Sequences	alignment bits	e-values
ref|YP_001331091.1|	502	1.00E-140
ref|NP_370636.1|	502	1.00E-140
ref|YP_415567.1|	500	5.00E-140
ref|YP_039579.1|	499	1.00E-139
ref|ZP_04016209.1|	498	1.00E-139
ref|ZP_03563936.1|	419	1.00E-115
ref|ZP_03986419.1|	245	3.00E-63
ref|ZP_03986418.1|	177	7.00E-43
ref|NP_373023.1|	177	7.00E-43
ref|YP_495072.1|	176	1.00E-42
ref|ZP_04827531.1|	176	2.00E-42
ref|YP_187302.1|	176	3.00E-42
ref|ZP_04864336.1|	160	1.00E-37
ref|ZP_04839552.1|	98.2	7.00E-19
ref|ZP_03561897.1|	83.6	2.00E-14
ref|ZP_04796428.1|	75.9	3.00E-12
ref|YP_002633364.1|	75.9	4.00E-12
ref|ZP_04824520.1|	75.5	4.00E-12

ref|NP_763945.1|	75.5	4.00E-12
ref|YP_254196.1|	75.1	5.00E-12
ref|ZP_04818422.1|	73.9	1.00E-11
ref|ZP_03613101.1|	73.9	1.00E-11
ref|ZP_04677773.1|	73.2	2.00E-11
ref|ZP_04059846.1|	72.8	3.00E-11
ref|YP_302193.1|	72.8	3.00E-11
ref|NP_371140.1|	71.2	9.00E-11
pdb|1FZP|D	71.2	9.00E-11
pdb|2FRH|A	70.9	1.00E-10
pdb|2FNP|A	69.7	2.00E-10
gb|AAB05396.1|	68.9	4.00E-10
gb|ABD73658.1|	65.5	5.00E-09
ref|ZP_03932304.1|	46.6	0.002
ref|ZP_03936223.1|	46.2	0.003
ref|ZP_04864335.1|	45.8	0.004
ref|ZP_04676978.1|	45.8	0.004
ref|ZP_05366829.1|	45.4	0.005
ref|ZP_03920664.1|	45.4	0.005
ref|ZP_05086994.1|	45.1	0.007
ref|YP_301092.1|	45.1	0.007
ref|ZP_03612535.1|	44.7	0.008
ref|YP_300706.1|	44.3	0.011
ref|ZP_04817518.1|	43.5	0.016
ref|YP_501258.1|	42.4	0.047
ref|YP_001576352.1|	42	0.053
ref|YP_187301.1|	42	0.053
ref|ZP_03957600.1|	41.6	0.074
ref|YP_002634059.1|	41.6	0.075
ref|YP_001231285.1|	41.6	0.076
ref|YP_188894.1|	41.6	0.079
ref|ZP_04797467.1|	41.2	0.082
ref|ZP_03931708.1|	41.2	0.083
ref|NP_764990.1|	41.2	0.085
ref|ZP_02431579.1|	41.2	0.087
ref|NP_373022.1|	41.2	0.097
ref|YP_001247875.1|	41.2	0.097
ref|ZP_04060580.1|	40.8	0.12
ref|YP_186645.1|	40.8	0.13
ref|YP_001353624.1|	40.4	0.15
ref|ZP_01304027.1|	40.4	0.16
ref|YP_147717.1|	40.4	0.16
ref|ZP_04819627.1|	40.4	0.16

ref|ZP_03752400.1|	40.4	0.17
ref|YP_001559761.1|	40.4	0.17
ref|NP_378388.1|	40	0.19
ref|YP_002634877.1|	40	0.21
ref|YP_002786464.1|	40	0.22
ref|NP_372288.1|	39.7	0.26
ref|ZP_01719933.1|	39.7	0.26
ref|YP_001332689.1|	39.7	0.29
ref|YP_043808.1|	39.7	0.29
ref|YP_494402.1|	39.7	0.3
ref|YP_041233.1|	39.7	0.31
ref|NP_391635.1|	39.3	0.33
ref|NP_694380.1|	38.9	0.41
ref|ZP_04370381.1|	38.9	0.43
ref|ZP_05367617.1|	38.9	0.43
ref|ZP_03613556.1|	38.9	0.46
ref|YP_002829202.1|	38.9	0.47
ref|YP_500382.1|	38.9	0.49
ref|ZP_05372472.1|	38.9	0.51
ref|YP_253074.1|	38.1	0.69
ref|YP_002831895.1|	38.1	0.72
ref|YP_002837327.1|	38.1	0.76
ref|ZP_04676980.1|	38.1	0.82
ref|ZP_04351651.1|	37.7	1
ref|YP_002561209.1|	37.7	1
ref|ZP_04767758.1|	37.7	1.1
ref|YP_083773.1|	37.7	1.1
ref|NP_342553.1|	37.7	1.1
ref|ZP_04798349.1|	37.4	1.3
ref|ZP_04826476.1|	37.4	1.3
ref|ZP_02171452.1|	37.4	1.3
ref|ZP_04803802.1|	37.4	1.3
ref|ZP_04059377.1|	37.4	1.3
ref|YP_189437.1|	37.4	1.5
ref|NP_765423.1|	37	1.6
ref|ZP_04798347.1|	37	1.6
ref|NP_765425.1|	37	1.7
pdb|3HRM|A	37	1.8
ref|ZP_03981559.1|	37	1.8

8.4 KNN for gene data analysis

Here we apply KNN to the *Burkholderia pseudomallei* gene expression data. The performance of the KNN model using the Manhattan distance is shown in the left panel of Fig. 8.9 and the performance of the KNN model using the Euclidean distance is shown in the right panel of Fig. 8.9. It can be seen that the best Manhattan model uses 3 nearest neighbours while the best Euclidean model uses 5 nearest neighbours.

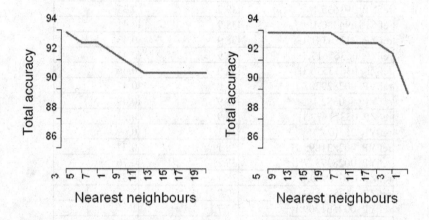

Fig. 8.9. The performance of KNN models for the *Burkholderia pseudomallei* gene expression data using the Euclidean distance (right panel) and the Manhattan distance (left panel). The horizontal axes indicate the number of nearest neighbours. The vertical axes indicate the total prediction accuracy.

Summary

This chapter has discussed two types of machine learning algorithms which are either simple in learning or simple in prediction. The discriminant analysis using Fisher algorithm (FDA) and the algorithm using nonlinear variables (QDA) are easy learning algorithms. The cost of learning only involves simple linear algebraic operations. The learned models can be easily interpreted, i.e. explaining which variables are important. KNN on the other hand has a very low learning cost but has the problem of prediction cost when the training data size is large. This chapter has also discussed the applications of these two types of algorithms in bioinformatics.

It must be noted that both FDA and QDA employ a linear learning procedure although QDA uses nonlinear variables. When the number of variables increases, the number of quadratic variables can increase dramatically. KNN, strictly speaking, is not a linear learning algorithm. When K is decreased it tends to be more nonlinear. When K is large the decision boundary is more smoothed out leading to a more linear classification property.

Chapter 9

Classification and Regression Trees, Random Forest Algorithm

This chapter discusses one typical supervised learning approach. This type of algorithms aims to mimic human-like decision-making systems for specific applications. The popularity of this type of algorithms results from its simplicity and intuitiveness. A model built this way enjoys some distinct features in interpreting predictions and displays data structures which are very welcome in many applications.

9.1 Introduction

The main objective of every prediction is for indicating what will happen in the future. The prediction that species diversification is positively related to species diversity [268] can warn us about how to maintain species diversity to reduce the risk of the extinction of many species which will finally affect human living. The prediction of protein modifications can narrow down the experiment targets. The prediction of cancer in its early stage may save or prolong life.

Many machine learning algorithms are able to deliver good prediction models through training. Compared with other machine learning algorithms, the inductive programming approach, including the decision tree algorithm [269, 270] and the classification and regression tree algorithm [271], has some distinct features suitable for bioinformatics projects. First, each prediction can be well interpreted. For instance, a cellular function prediction model can indicate which genotypic variable is the factor initiating a specific cellular function. Second, both numerical data and categorical data can be modelled. This is particularly important in analysing biological data where categorical data often occurs. Third, these algorithms need much shorter model construction time compared with other machine learning algorithms. Fourth, they are capable of handling data with very large dimensionality. Many other machine learning algorithms need the number of input vectors to be larger than the number of model parameters to ensure statistical learning significance. Finally the outcome of each learning process will deliver a clear data structure demonstrating the hidden knowledge underlying data.

9.2 Basic principle for constructing a classification tree

The basic principle of the algorithms discussed in this chapter is "divide and conquer". Using this principle, a data space with mixed classes of input vectors is divided into two sub-spaces using classification rule.

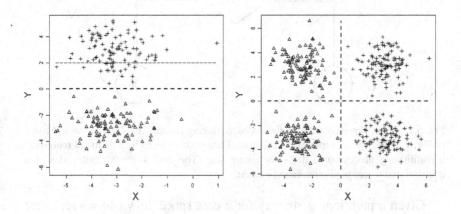

Fig. 9.1. An illustration of dividing a data space. Triangles and crosses represent two different classes of data. The long broken lines are used for partitioning the spaces. X and Y are two variables. The thin dotted line in the left panel is an alternative decision boundary for dividing the space.

For instance, in the left panel of Fig. 9.1, one decision boundary (y = 0) can be used to partition the space into two sub-spaces in which input vectors can be easily classified. However, the partitioning using the decision boundary (y = 2) cannot make two sub-spaces in which input vectors can be well-separated. On the right panel of Fig. 9.1, a further decision boundary (x = 0) is added to make four sub-spaces for classification of input vectors. After a data space is well-divided into small sub-spaces in which input vectors can be well classified, the divided sub-spaces can be used to develop classification rules. For instance, if a novel input vector is found in the upper sub-space in the left panel, this novel input vector is then labelled by the input vectors in this upper sub-space.

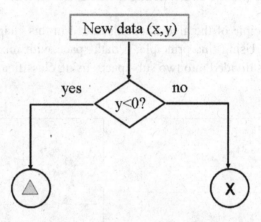

Fig. 9.2. A tree representation of the decision-making process involved in the left panel of Fig. 9.1. The square represents data input. Each node denoted by a diamond represents a partitioning process or a decision-making step. The circles are the end nodes of a decision tree where predictions can be made.

Given a partitioning strategy for a data space, how can we represent this partitioning space as a decision-making process? In order to explore human intelligence of a model, a tree-like decision-making structure has been adopted in the decision tree algorithm [269, 270] and the classification regression tree algorithm [271]. For instance, the decision making process in the left panel of Fig. 9.1 can be expressed as a tree shown in Fig. 9.2 while the decision-making process in the right panel of Fig. 9.1 can be demonstrated in Fig. 9.3.

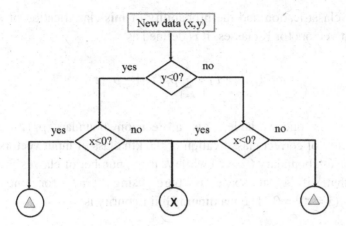

Fig. 9.3. A tree representation of the decision-making process involved in the right panel of Fig. 9.1. For the explanation of the representation refer to Fig. 9.2.

To make a best data space partitioning and deliver the best decision-making system or model, we need to consider how to find the best decision boundary to partition a space, i.e. how to make a root and branch node shown in Fig. 9.2 and Fig. 9.3. For instance, the thin dotted line in the left panel of Fig. 9.1 is certainly not a good choice although the majority of input vectors can be separated. Intuitively, we can say that the decision boundary (y = 0) represented by the long broken line in the left panel of Fig. 9.1 is a good one while the decision boundary (y = 2) marked by the dotted line is not a good candidate. This is because both sub-spaces generated by the decision boundary (y = 0) are *pure* for one class of input vectors. However, the upper sub-space generated by the decision boundary (y = 2) is *pure* while the lower sub-space is *impure*. To automate the selection of the best decision boundary, a quantitative measure for purity or impurity is required. Two ways to measure the impurity of a node in a tree have been proposed. One is called Gini impurity and the other is called information gain.

Gini impurity has been used for selecting the decision boundary in various algorithms such as ID3, C4.5, C5, classification and regression tree (CART) and the random forest algorithm [269-272]. The Gini impurity is calculated by summing the products of the probability of correct classification and the probability of mis-classification of a class of input vectors for K classes. It is defined as

$$I_G(x = \tau) = \sum_{k=1}^{K} p_k(\tau)[1 - p_k(\tau)] \tag{9.1}$$

where x is one variable, τ is a decision boundary, $p_k(\tau)$ is the probability of correct classification of the kth class of input vectors using the decision boundary ($x = \tau$) while K is the number of classes. It can be seen that, if a sub-space is pure using $x = \tau$ for one class, $\min\{I_G(x = \tau)\} = 0$. The maximum Gini impurity is

$$\max\{I_G(x = \tau)\} = \sum_{k=1}^{K} \frac{1}{K} \frac{K-1}{K} = \frac{K-1}{K} \tag{9.2}$$

The other measure is called information gain based on entropy which is defined as

$$I_E(x = \tau) = -\sum_{k=1}^{K} p_k(\tau) \log_2 p_k(\tau) \tag{9.3}$$

When a sub-space is pure for one class, $\min\{I_E(x = \tau)\} = 0$. The more the classes of input vectors are in a sub-space, the larger the entropy is. The largest information gain is

$$\max\{I_E(x = \tau)\} = -\sum_{k=1}^{K} \frac{1}{K} \log_2 \frac{1}{K} = \log_2 K \tag{9.4}$$

The construction of a decision tree or classification and regression tree model is based on repeated optimisation of one of the impurity mentioned above.

In the sections below, we discuss two typical algorithms. One is called the classification and regression tree algorithm (CART) and the other is called the random forest algorithm (RF).

9.3 Classification and regression tree

CART can be used for both classification analysis and regression analysis and makes a prediction model in three steps described as below.

Step 1: tree growing. For a given data set, a tree is grown based on recursive partitioning of the data space using decision boundaries. For every partition using a decision boundary, a node is formed. The node is composed of two parts. One is the variable selected (such as x) and the other is the threshold (such as T). The given data space or a sub-space is divided into two sub-spaces according to the relationship between the x value and T value for all input vectors. The impurity is calculated for this new node. If the impurity is zero, no further partition is taken beyond this node. In this situation the nodes below this node are labelled according to the class property of the input vectors in the sub-spaces.

Step 2: tree pruning. A tree is pruned if it fails to generate better prediction performance compared to a tree with a simpler structure. Note that when pruning is completed, a leaf node may not be pure for one class of input vectors. In this case, a probability of belonging to one class is calculated according to the fraction of one class of data points over total data points.

Step 3: tree selection. An optimal tree is selected if it outperforms the other candidates in predicting novel data.

After construction, a CART model can be used for predictions. As indicated in Figures 9.2 and 9.3, a novel input vector is fed into the model. The root node examines the relevant variable's value against the threshold to determine if the action moves to the left sub-tree or the right sub-tree. The same examination applies to all the following branch nodes in the tree. When a leaf node is reached, the prediction is made according to the maximum posterior probability. The posterior probability is defined in chapter 7. In a CART model, the conditional probability of the kth class (given K classes) is defined as the fraction of input vectors belonging to the kth class in a sub-space associated with a leaf node.

CART has been used for classifying substrates, inhibitors, and inducers of p-glycoprotein [273], for identifying head and neck squamous cell carcinoma [274], for detecting SNP-SNP interaction

[275], for HIV-I drug (CCL3L1-CCR5) evaluation [276], for studying the relation between genetic polymorphisms in double-strand break DNA repair genes and oral premalignant lesions [277], for studying how genetic variants in cell cycle can control pathways related to susceptibility of bladder cancer [278], for detecting breast cancer using genomic data [279], and for studying mRNA expression data variance [280].

9.4 CART for compound pathway involvement prediction

In this section we study the prediction of pathway involvement of a compound. Compounds are downloaded from the KEGG library [155]. In this data set, 14423 compounds are found among which 2961 compounds have metabolic pathway annotations. Among them, 1050 are in biosynthesis pathways, 501 are in degradation pathways, and 1491 are in metabolism pathways comprising both biosynthesis and degradation.

Each compound is represented by a formula in KEGG and is composed of chemical elements and their quantities. For instance, the compounds Cyanate, Carbamate, and Urea are represented by $C_1H_1N_1O_1$, $C_1H_2N_1O_2$, and $C_1H_4N_2O_1$, respectively. H means Hydrogen, N means Nitrogen and O means Oxygen. The numbers represent the quantities of the chemical elements. In both databases, one formula can be shared by multiple compounds. This is because a formula reflects how chemical elements are contained in a compound, but does not fully illustrate compound structure.

In order to have an unbiased evaluation of model performance, all the duplicated formulae are carefully examined. If duplicated formulae have different compound names in the same pathway category, only one of them is kept for the study. If duplicated formulae have different compound names in different pathways, all are removed. By this examination, 382 compounds in biosyntheses pathways, 155 compounds in degradation pathways and 501 compounds in metabolism pathways are retained for the study. We then build a model to map compounds into these three pathway categories made by KEGG using information stored in compounds' formulae. The mass values of chemical elements (http://www.wsearch.com.au/) [281] are used for encoding compounds (formulae). The formula of each compound is encoded according to the presence of chemical elements as well as the quantities of them. For instance, the Hydrogen in three compounds mentioned above can be encoded by 1.0078250321, 2*1.0078250321 and 4*1.0078250321, respectively, whilst the Oxygen can be encoded by 15.9949146221, 2*15.9949146221, and 15.9949146221, respectively. The use of chemical element weights for encoding compounds can represent compounds' chemical element property well.

In machine learning, an attribute (a chemical element in this context) with a small occurrence rate normally will not make a significant contribution to model performance. Because of this, chemical elements with <1% occurrence rates in our data are dropped. This filtering process leads to seven most contributing chemical elements. They are Hydrogen (gas), Carbon (solid), Nitrogen (gas), Oxygen (gas), Phosphorus (solid), Sulphur (solid), and Chlorine (gas). Each compound is then encoded using these seven chemical elements. Figure 9.4 shows the classification tree for this data. In the Figure, "0" means biosynthesis, "1" means

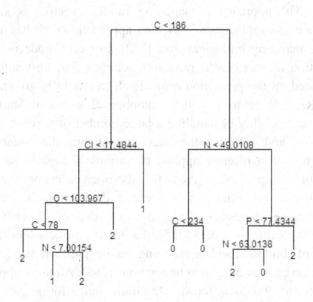

Fig. 9.4. A classification tree generated for the prediction of metabolic pathway involvement of compounds.

degradation, and "2" means metabolism. In the tree, it has been found that biosynthesis pathway is classified by the following factors at the terminal nodes: 1) Oxygen weight > 7.997462; 2) Oxygen weight > 55.9822; 3) Hydrogen weight > 21.6682; 4) Nitrogen weight > 63.0138; 5) Carbon>318. However, metabolism is not classified using Hydrogen. Instead it uses: 1) Carbon weight < 66; 2) Carbon weight > 78; 3) Oxygen weight < 55.9822; 4) Nitrogen weight < 63.0138; and 5) Phosphorus weight > 77.4344. The classification of degradation pathway involvement is based on: 1) Chlorine weight > 17.4844; 2) Carbon < 78, and 3) Oxygen weight < 7.997462.

9.5 The random forest algorithm

The random forest (RF) algorithm is an extension to CART. RF is a newly developed machine learning algorithm [272]. The basic idea is to construct many trees using random vectors sampled from a data set. For the kth tree, a random vector is generated independently from the random vectors generated for the past k-1 trees. The remaining data are used for prediction. The approach of sampling random vectors is similar to bootstrap, i.e. the replacement sampling approach, which has also been applied to analysing biological data [282]. For each node in a tree, a small fraction of variables is randomly selected. The best split for the node is based on the prediction error. Each tree is fully grown without pruning. RF is able to provide a number of excellent features, for instance, the capability of handling a large number of variables, ranking the variables, and detecting the interaction among the variables. The algorithm has been recently applied to various biological data mining projects, for example, the prediction of the interactions between HIV-1 and human proteins using gene expression data [283], the analysis of differential gene expression [284], the diagnosis of ulcerative colitis based on gene expression data [285], the detection of cancers [286], the prediction of childhood leukaemia using gene expression data [287], and the prediction of protein-protein interactions [288]. All these applications show that the random forest algorithm outperforms some other algorithms.

9.6 RF for analysing *Burkholderia pseudomallei* gene expression profiles

The following case is an application of RF to the *Burkholderia pseudomallei* gene expression profile data. Refer to previous chapters for a description of this data. We first analyse the feature provided by RF in ranking variables (genes in this study). RF can provide two measures for ranking variables, one being the mean decrease in Gini gain and the other being the mean decrease in accuracy. The mean decrease in Gini gain is used to measure the quality of a split for each variable in a tree. Whenever the Gini gain of a node's descent nodes is less than the node's Gini gain, a split is carried out. The decrease of the Gini gain of this node is recorded. A variable such as a residue code or a residue correlation code may be used by different nodes for splitting or tree growing. It may therefore have different decreases of Gini gain at different nodes. A mean decrease in Gini gain across all nodes using a variable is calculated to measure the importance of the variable. The mean decrease in accuracy is calculated in a similar way by examining all nodes using the same variable. The mean decrease in Gini is shown in Fig. 9.5 and the mean decrease in accuracy is shown in Fig. 9.6. The top ten genes selected by two ranking criteria are shown in Table 9.1. The top ten are identical but have different orders.

Table 9.1. The top ten genes selected by mean decrease in Gini and mean decrease in accuracy for the *Burkholderia pseudomallei* gene data using RF.

	Mean decrease Gini	Mean decrease accuracy
1	BPSL2697	BPSL2697
2	BPSL2522	BPSS1512
3	BPSS1512	BPSL2522
4	BPSS0477	BPSL2096
5	BPSL2096	BPSS0477
6	BPSS1525	BPSS1532.1
7	BPSL2520	BPSS0476
8	BPSS0476	BPSS1532
9	BPSS1532	BPSL2520
10	BPSS1532.1	BPSS1525

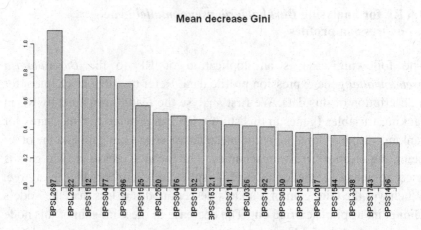

Fig. 9.5. The mean decrease in Gini of the top 20 genes selected by the RF model built for the *Burkholderia pseudomallei* gene expression data.

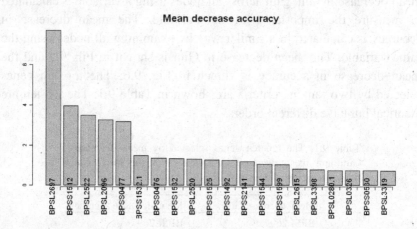

Fig. 9.6. The mean decrease in accuracy of the top 20 genes selected by the RF model built for the *Burkholderia pseudomallei* gene expression data.

Five-fold cross-validation is used for model evaluation. The predictions are analysed using both density analysis and ROC analysis. Figure 9.7 shows the density analysis and the ROC analysis. The prediction specificity is 95%. The sensitivity is 93%. The total accuracy is 95% and the area under ROC curve (AUR) is 0.96. The density shows clearly two separated clusters of the predictions.

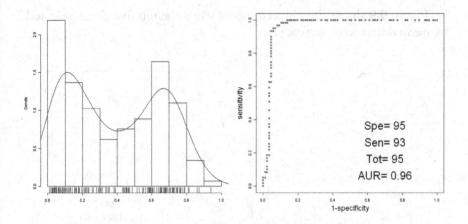

Fig. 9.7. The density analysis (left panel) and the ROC analysis (right panel) of predictions from five 5-fold cross-validation RF models for the *Burkholderia pseudomallei* gene data. In the ROC curve, the horizontal axes represent 1 – specificity and the vertical axes represent sensitivity.

We then use the top five genes selected by the mean decrease in Gini gain and mean decrease in accuracy ranking criteria. Figure 9.8 shows the density analysis and the ROC analysis of the reduced model using the top five genes generated by the mean decrease in Gini gain.

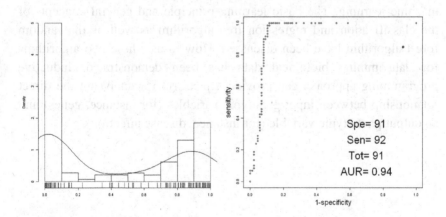

Fig. 9.8. Density (left panel) and ROC (right panel) analyses of the reduced model using the top five genes generated by mean decrease in Gini gain. In the ROC curve, the horizontal axes represent 1 – specificity and the vertical axes represent sensitivity.

Figure 9.9 shows the reduced model using the top five genes selected by mean decrease in accuracy.

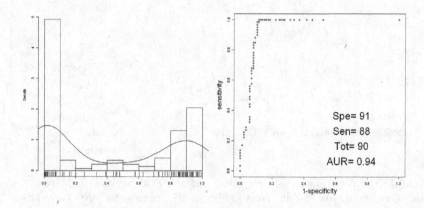

Fig. 9.9. Density (left panel) and ROC (right panel) analyses of the reduced model using the top five genes generated by mean decrease in accuracy. In the ROC curve, the horizontal axes represent 1 – specificity and the vertical axes represent sensitivity.

Summary

This chapter has discussed an inductive programming approach in machine learning. The basic learning principle and general concepts of the classification and regression tree algorithm as well as the random forest algorithm have been discussed. How to use these two algorithms for data mining biological data has been demonstrated. Inductive programming approach can provide a platform for analysing the direct relationship between input genotypic variables (for instance, genes) and an output phenotypic variable (for instance, disease infection).

Chapter 10

Multi-layer Perceptron

This chapter mainly discusses multi-layer perceptron (MLP) for supervised learning. Compared with many other nonlinear artificial neural learning algorithms, it has the advantages of modelling arbitrary nonlinear data. MLP has been widely used in many applications including bioinformatics. This chapter will focus on the history of MLP, the structure of MLP, the learning algorithm of MLP and its applications to bioinformatics.

10.1 Introduction

Neural networks are a class of computational algorithms mimicking the human brain with the support of modern, fast and sometimes parallel computational facility. In terms of this, neural networks are regarded as a class of information processing systems as well. The interpretation of this is that neural networks can re-construct an unknown function using the available data without any prior knowledge about function structures and parameters. Information processing has two meanings. The first is that neural networks can help to estimate function structures and parameters without domain experts involved. This is perhaps the most important reason for neural networks being so popular in many areas. The second is that neural networks are a class of intelligent learning machines which can store knowledge through learning as the human brain does for pattern recognition, decision making, novelty detection and prediction. Combining these two important factors, neural networks then become a powerful computational approach for handling data for various learning problems.

133

Neural network studies and applications have experienced several important stages. In the early days, neural network studies only focused on theoretical subjects, i.e. investigating if a machine can replace a human for decision-making and pattern recognition. The pioneer researchers were Warren McCulloch and Walter Pitts; [289] they showed the possibility of constructing a net of neurons which can interact with each other. The net was based on symbolic logic relationships. Table 10.1 shows one of McCulloch and Pitts OR logic, where the output is a logic OR function of two inputs.

Table 10.1. McCulloch and Pitts OR logic.

Input 1	Input 2	Output
0	0	0
0	1	1
1	0	1
1	1	1

This earlier idea of McCulloch and Pitts was not based on rigorous development as indicated by Fitch [290] that "in any case there is no rigorous construction of a logical calculus". However, the study on neural networks was continuing. For instance, Hebb in his book published in 1949 gave the evidence that McCulloch-Pitts model certainly works [291]. He showed how neural pathways can be strengthened whenever they are activated. In his book, he indicated that "when an axon of cell A is near enough to excite a cell B and repeatedly or persistently takes part in firing it, some growth process or metabolite change takes place in one or both cells such that A's efficiency, as one of the cells firing B, is increased". In 1954, Marvin Minsky completed his doctorial study on neural networks. His dissertation was entitled "Theory of Neural-Analog Reinforcement Systems and its Application to Brain-Model Problem". Later he published a paper about this work in a book [292]. This triggered a wide scale of neural network research. In 1958, Frank Rosenblatt built a computer at Cornell University called the Perceptron (later being called single-layer perceptron) which can learn new skills by trial and error through mimicking the human thought process. However, this work was evaluated by Minsky in 1969 [293] showing its incapability in dealing with complicated data. Minsky's book then blocked the further study of neural networks for many years.

In the period of the 1970's and 1980's, neural network research was in fact not completely ceased. For instance, the self-organizing map [198] and the Hopfiled net were intensively studied [294]. In 1974, Paul Werbos conducted his doctorial study at Harvard University and studied the training process called back propagation of errors. The work was published later in his book [295]. This important contribution led to the work of David Rumelhart and his colleagues in the 1980's. In 1986, the back propagation algorithm was introduced by Rumelhart and his colleagues with the implementation called the delta rule for supervised learning problems [296]. The neural net structure is called multi-layer perceptron (MLP). Since development, MLP became very popular for data mining or machine learning in both theoretical studies and practical exercises.

The most important contribution of Rumelhart and his colleagues' work is that a simple training or learning algorithm based on trial-and-error principle has been implemented and has demonstrated its powerfulness in dealing with problems which were declared impossible by Minsky in 1969. In contrast to Rosenblatt's single-layer perceptron (SLP), the most important difference in MLP is the introduction of hidden neurons.

Shown in Fig. 10.1 (a) is a structure of SLP, where there are three input neurons named as x1, x2, and x3 with a single output neuron named as y. In contrast, a structure of MLP is shown on the right panel in Fig. 10.1, where in addition to three input neurons and an output neuron, three hidden neurons named as z1, z2, and z3 are inserted between the input and output neurons. Generally, x1, x2, and x3 represent observed values for three independent variables (or the input variables) while y corresponds to observed values for a dependent variable (or the output variable). The hidden neurons represent variables which are not observed.

Practically, there are three related subjects in using neural networks for any applications. They are model construction, model selection, and model evaluation. In addition to these three practical issues, we need to be aware of three theoretical issues: parameter estimate, learning rule, and learning algorithms. Parameter estimate is a learning process by which knowledge in data is extracted and expressed quantitatively in a neural network model. The extracted knowledge is ultimately used for making predictions about unseen data. In most cases, we have no idea what values should be assigned to model parameters when we have data only. The data obtained are the only source for us to estimate model parameters. Besides we need to determine an optimized model structure to represent true knowledge hidden in data. For different supervised learning projects, different learning algorithms are needed. Moreover, there might be many variants in one type of supervised learning. There are various learning rules available. Some are based on numerical methods and some are based on statistical approaches. Some are fast for some types of data and some are accurate for some types of data. We will focus on numerical approaches for deriving neural network learning rules.

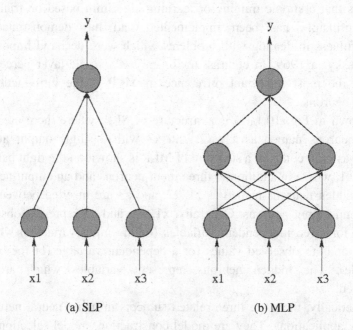

(a) SLP (b) MLP

Fig. 10.1 SLP and MLP structure.

10.2 Learning theory

10.2.1 *Parameterization of a neural network*

A neural network without parameters will have no capability of associative memory. In particular, a neural network whenever its structure has been determined must possess the power for prediction in a supervised learning project. In order to make a neural network capable of prediction, it must have parameters which represent processed information. This can be explained by a simple example. Suppose we are interested in studying whether a metabolite in a specific pathway is related to its upstream metabolite. We first denote this metabolite as y. Meanwhile, we denote three upstream metabolites for the downstream metabolite y as x_1, x_2, and x_3. Suppose we have had some observations for x_1, x_2, x_3, and y. Our objective is to construct a model which can establish the relationship between x_1, x_2, x_3, and y as a predictive function $y = f(x_1, x_2, x_3)$. If $f(x_1, x_2, x_3)$ is properly parameterized, say $y = f(w_0 + w_1 x_1 + w_2 x_2 + w_3 x_3)$, where w_0 is a bias term, and w_1, w_2, and w_3 are parameters for the three upstream metabolites, we can make a prediction whenever we have new values for x_1, x_2, and x_3.

It is normally believed that parameters in a neural network model represent the knowledge in data. For instance, if a neural network model is expressed as $y = \sigma(0.1 + 0.03x_1 + 5.1x_2 + 0.002x_3)$, all three input variables have the same magnitude and $\sigma(z)$ is a monotonic linear function of z, i.e. $\sigma(z) \propto z$, we can believe that x_2 plays a key role for y. In other words, x_2 is the dominant factor for y. Ignoring the other two input variables will not lose much precision in prediction.

10.2.2 *Learning rules*

Before discussing learning rules, we need to establish a proper objective function. There are normally two types of objective functions for supervised neural learning. They are the square error function and the cross-entropy function. The former is used for regression analysis which addresses a type of problems of continuous function approximation. The latter is used for classification analysis which addresses a class of applications of data partitioning.

We normally denote a regression function as

$$y_n = f(\mathbf{x}_n, \mathbf{w}) \tag{10.1}$$

Here $\mathbf{w} \in \mathfrak{R}^H$ is a numerical parameter vector of H dimensions and $\mathbf{x}_n \in \mathfrak{R}^D$ is a numerical input vector of D dimensions describing the nth object in a data set, where $\mathfrak{R}$ is the real number set. $H > D$. Correspondingly, $y_n \in \mathfrak{R}$ is the model output for $\mathbf{x}_n$. H is heavily dependent on a model's structure. For $\mathbf{x}_n$, we normally have its observed phenotypic property called target $t_n \in \mathfrak{R}$. Note that t_n does not represent a true value in most cases. Normally, it is called a corrupted function value. For instance, a true function is a sin function $5\sin(x)$. We may have observed corrupted values from a noise-added sin function $5\sin(x) + G(0,1)$, where $G(0,1)$ is called a white noise. The existence of noise is normally unavoidable in many experiments. Many factors can result in noise. In order to estimate the parameter vector $\mathbf{w}$, we need to make the distance between y_n and t_n $(t_n - y_n)$ as small as possible during learning. Based on this, we have a commonly used square error function (mean square error function) for regression analysis as below

$$\varepsilon = \frac{1}{N} \sum_{n=1}^{N} (t_n - y_n)^2 \tag{10.2}$$

Here N is the number of observed pairs $(\mathbf{x}_n, t_n)$. A learning rule must ensure that the model parameter vector satisfies

$$\tilde{\mathbf{w}} = \min_{\text{arg}} \left\{ \frac{1}{N} \sum_{n=1}^{N} (t_n - f(\mathbf{x}_n, \hat{\mathbf{w}}))^2 \right\} \qquad \forall \hat{\mathbf{w}} \in \mathfrak{R}^H \tag{10.3}$$

Here $\hat{\mathbf{w}}$ is a vector (a point) in an H-dimensional space (called a parameter space) and $\tilde{\mathbf{w}}$ is the optimal vector among many (normally infinite) $\hat{\mathbf{w}}$'s.

In classification, a different objective function is used if the model output y_n is constrained in the interval $[0,1]$. Neural networks employing the sigmoid function can easily fulfil this requirement. The cross-entropy function is normally employed for discriminant analysis, where $t_n \in \{0,1\}$

$$O = \prod_{n=1}^{N} y_n^{t_n} (1 - y_n)^{1-t_n} \tag{10.4}$$

In most cases, negative logarithm is applied to this objective function leading to

$$O = -\sum_{n=1}^{N} t_n \log y_n + (1 - t_n) \log(1 - y_n) \tag{10.5}$$

A learning process aims to minimize this objective function so that

$$\tilde{\mathbf{w}} = \min_{\arg} \left\{ -\sum_{n=1}^{N} t_n \log f(\mathbf{x}_n, \hat{\mathbf{w}}) + (1 - t_n) \log(1 - f(\mathbf{x}_n, \hat{\mathbf{w}})) \right\} \tag{10.6}$$

$$\forall \hat{\mathbf{w}} \in \Re^H$$

It is then obvious that we have to analyse the function $f(\mathbf{x}, \mathbf{w})$ before discussing the learning rule. In neural networks, the sigmoid function is normally used for $f(\mathbf{x}, \mathbf{w})$ because it has two advantages, i.e. being derivable and parallelism. The former makes it possible to apply conventional numerical approximation approaches which heavily depend on derivatives to parameter estimation and the latter makes it possible to use parallel computing techniques because the calculation of each neuron output is completely independent from the calculations of other neuron's outputs in the same layer. The sigmoid function is defined as below

$$f(z) = \frac{1}{1 + \exp(-z)} \tag{10.7}$$

It is not very difficult to see that the sigmoid function squashes the value of z in to the interval $(0,1)$ as

$$\lim_{z \to -\infty} \frac{1}{1 + \exp(-z)} = 0 \qquad (10.8)$$

and

$$\lim_{z \to +\infty} \frac{1}{1 + \exp(-z)} = 1 \qquad (10.9)$$

In addition, the other advantage of the sigmoid function is that its derivative is easily calculated as the entropy as below

$$\frac{df(z)}{dz} = f(z)(1 - f(z)) \qquad (10.10)$$

We now use regression analysis as an example for the analysis of the learning rule. In most cases, we will have no knowledge of what values should be assigned to model parameters. Like statistical learning, neural learning also starts from a random guess, i.e. assigning random values to model parameters (called initialization) and based on these random parameters, we start to search for the way by which an objective function can be decreased, hence bringing the current model parameters ($\hat{\mathbf{w}}$) closer to the optimal solution ($\tilde{\mathbf{w}}$). As we know, regression analysis adopts a quadratic-like objective (error) function. In a quadratic function, there will always be some relationship between the derivatives and the optimal solution. Figure 10.2 shows such a relationship for a case where there is only one model parameter. Two filled dots are the possible random guesses. It can be seen that the optimal model parameter must correspond to the bottom of the valley of the quadratic objective function. The slope (first derivative) denoted by a straight line of the random guess on the left side of the optimal solution shows a negative sign while the slope denoted by another straight line of the random guess on the right side of the optimal solution shows a positive sign. The negative sign means that when the value of w increases, the error O decreases. The positive sign means that when the value of w increases, the error O increases. From this, it can be seen that we must increase the model parameter when the slope of the model output based on the current model parameter shows a negative sign. We must decrease the model parameter when the slope of the model output based on the current model parameter shows a positive sign. The slope of model output is mathematically defined as the first derivative of the objective function with respect to the model parameter as below

$$\nabla O = \frac{dO}{dw} \qquad (10.11)$$

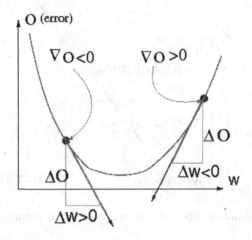

Fig. 10.2. The relationship between a model output using the current model parameter and the direction of the optimal model parameter.

Before defining the quantitative learning rule which will be used to update model parameters stochastically, we need to analyze the qualitative relationship between parameter change and the slope. The next thing is to determine the learning rule quantitatively.

If the change (increase or decrease) of w is denoted by Δw, we then have a qualitative relationship from Fig. 10.3 that if the absolute value of the slope is larger, the current position is more departed from the optimal solution and if the absolute value of the slope is smaller, the current position is closer to the optimal solution. When w is closer to the optimal solution, we must make a smaller change to w so that we will not miss the optimal solution. When w is more departed from the optimal solution, we can have a larger change of w. From this, we then have a qualitative learning rule defined as below

$$| \Delta w | \propto | \nabla O |$$ (10.12)

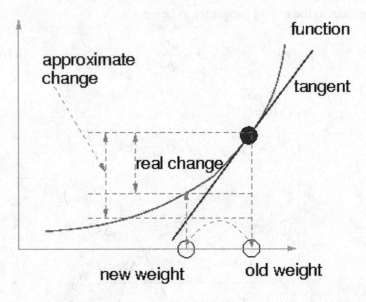

Fig. 10.3. The quantitative relationship between slope and the magnitude of model parameter change.

Quantitatively, the learning rule (also called the delta rule) is defined as below

$$\Delta w = -\eta \nabla O$$ (10.13)

Here $\eta \in (0,1)$ is called the learning rate. The delta rule may not always work properly. It is quite often that a new solution of w may miss the optimal solution. For instance, the new solution w_1 generated using the delta rule from w_0 misses the optimal solution, i.e. the valley of the quadratic curve as seen in Fig. 10.4. From w_1 the delta rule will lead to w_2^A which again misses the optimal solution. However, we have noticed that the first derivatives at w_0 and w_1 have different signs meaning that they have a complementary function. If the first derivative at w_{t+1} has a different sign to the one at w_t, it means that w_t and w_{t+1} are sitting on opposite sides of the optimal solution, see Fig. 10.2. The move from w_{t+1} to w_{t+2} may miss the optimal solution again. If we can correct the move from w_{t+1} to w_{t+2} using a momentum factor which has a different first derivative sign to the one at w_{t+1}, the risk can possibly be reduced. Remember that the first derivative at w_t is different to the one at w_{t+1}. We can design a revised delta rule for this purpose

$$\Delta w^{t+1} = -\eta \nabla O^t + \alpha \Delta w^t \qquad (10.14)$$

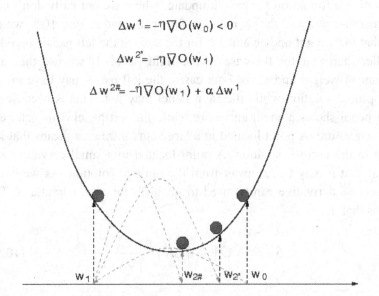

Fig. 10.4. The illustration of the use of the momentum factor for fast learning.

Here Δw^{t+1} is the update of w at time $t+1$, Δw^t is the update of w at time t, ∇O^t is the first derivative of O with respect to w at time t and $\alpha \in (0,1)$ is a positive number called the momentum factor. In Fig. 10.4, we can see that this revised delta rule can reduce this risk. This time, the move from w_1 is to w_2^B rather than w_2^A. According to equation (10.14), we can see that

$$\Delta w^{t+1} = -\eta(\nabla O^t + \alpha \nabla O^{t-1}) + \alpha^2 \Delta w^{t-1} \qquad (10.15)$$

From the above equation, we can conclude two aspects. First, if ∇O^t and ∇O^{t-1} have the same sign, the previous update instruction (∇O^{t-1}) will enhance the new update instruction (∇O^t), otherwise ∇O^{t-1} will reduce the impact of ∇O^t. Second, if Δw^{t+1} and Δw^{t-1} have the same sign, Δw^{t-1} will enhance Δw^{t+1}. Otherwise, Δw^{t-1} will reduce the impact of Δw^{t+1}.

In using the delta rule or the revised delta rule, the user needs to tune the learning rate and the momentum factor to proper values. This is not an easy job. There is another numerical method which uses second derivative information for weight update, where we normally don't need the learning rate and the momentum factor. Shown in Fig. 10.5, we can see that the weight update amount for the case in the left panel should be smaller than that for the case in the right panel. If we use the same amount of weight update for both cases, the left panel may have missed the optimal solution while the right panel may not. This is because the right panel shows a small curvature while the left panel demonstrates a large curvature. A point located in a large curvature area means that it is close to the optimal solution. A point located in a small curvature area means that it may be far away from the optimal solution. As we know, the second derivative can be used to quantify function curvatures. This means that

$$| \Delta w | \propto | \nabla O | \qquad | \Delta w | \propto \frac{1}{| \nabla \nabla O |} \qquad (10.16)$$

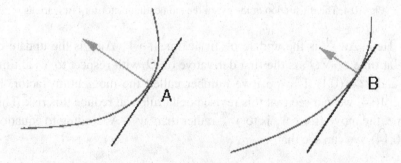

Fig. 10.5. The illustration of using second derivative information for weight update.

Here ∇O and $\nabla\nabla O$ are the first and second derivatives with respect to w. The update rule using the second derivative information is called the Newton-Raphson method. In application to neural network parameters, it is illustrated as below

$$\Delta w = -\frac{\nabla O}{\nabla\nabla O} \tag{10.17}$$

or

$$\Delta \mathbf{w} = -\mathbf{H}^{-1}\nabla O \tag{10.18}$$

Here $\mathbf{w}$ is a weight vector and $\mathbf{H}$ is called a Hessian matrix of second derivatives as below

$$\mathbf{H} = \begin{pmatrix} \dfrac{\partial^2 O}{\partial w_1 \partial w_1} & \dfrac{\partial^2 O}{\partial w_1 \partial w_2} & \cdots & \dfrac{\partial^2 O}{\partial w_1 \partial w_m} \\ \dfrac{\partial^2 O}{\partial w_2 \partial w_1} & \dfrac{\partial^2 O}{\partial w_2 \partial w_2} & \cdots & \dfrac{\partial^2 O}{\partial w_2 \partial w_m} \\ \vdots & \vdots & \vdots & \vdots \\ \dfrac{\partial^2 O}{\partial w_m \partial w_1} & \dfrac{\partial^2 O}{\partial w_m \partial w_2} & \cdots & \dfrac{\partial^2 O}{\partial w_m \partial w_m} \end{pmatrix} \tag{10.19}$$

where $\dfrac{\partial^2 O}{\partial w_i \partial w_j}$ is the second derivative of O with respect to w_i and w_j.

10.3 Learning algorithms

In this sub-section, we discuss two learning algorithms for regression and classification analyses respectively, where different objective functions are used.

10.3.1 *Regression*

In regression analysis, the target variable is commonly a numerical variable $t_n \in \Re$ (or $t_n \in [0,1]$). In this case, the least mean square error function is used as the objective function as seen in equation (10.2). Using the revised delta rule (equation 10.14), we then have two update rules as below. First, the update rule for the weights between hidden neurons and an output neuron (for instance, between hidden neurons z_1, z_2, z_3 and output neuron y in Fig. 10.1) is

$$\Delta\mathbf{w}_0^{t+1} = \eta\mathbf{Z}^T\mathbf{B}\mathbf{e} + \alpha\Delta\mathbf{w}_0^t \qquad (10.20)$$

Here $\mathbf{w}_0 = (w_{01}, w_{02}, \cdots, w_{0H})^T$ is the hidden weight vector with w_{0h} connecting the hth hidden neuron to the output neuron, $\mathbf{e} = (e_1, e_2, \cdots, e_\ell)^T$ is the error vector with $e_n = t_n - y_n$, $\mathbf{B} = diag\{y_n(1-y_n)\}$ is the diagonal entropy matrix of outputs with N rows and N columns, and $\mathbf{Z}$ is the matrix recording the outputs from all the hidden neurons with N rows and H columns (H hidden neurons). Second, the update rule for the weights between input neurons and the hth hidden neuron (for instance, between the input neurons and the hidden neuron z_1 in Fig. 10.1) is shown as below

$$\Delta\mathbf{w}_h^{t+1} = w_{0h}\eta\mathbf{X}^T\mathbf{B}\mathbf{Q}_h\mathbf{e} + \alpha\Delta\mathbf{w}_h^t \qquad (10.21)$$

Here $\mathbf{w}_h = (w_{h1}, w_{h2}, \cdots, w_{hD})^T$ is the input weight vector with w_{hd} connecting the hth hidden neuron to the dth input neuron, $\mathbf{Q}_h = diag\{z_{nh}(1-z_{nh})\}$ is the diagonal entropy matrix for the hth hidden neuron with N rows and N columns, and $\mathbf{X}$ is the matrix recording all the input vectors, i.e. having N rows and D columns (N input vectors and D input variables).

10.3.2 *Classification*

For a classification problem, the target variable is commonly a discrete variable $t_n \in I$ with I meaning integers. We study discrimination problems where $t_n \in \{0,1\}$ in this chapter. The cross-entropy function is commonly used as the objective function for classification projects as seen in equation (10.4). Applying the revised delta rule to equation (10.5), we will also have two update rules. First, the update rule for the weights between hidden neurons and the output neuron if we have one output neuron (for instance, between hidden neurons z_1, z_2, z_3 and the output neuron y in Fig. 10.1) is

$$\Delta \mathbf{w}_0^{t+1} = \eta \mathbf{Z}^T \mathbf{e} + \alpha \Delta \mathbf{w}_0^t \tag{10.22}$$

Second, the update rule for the weights between input neurons and the hth hidden neuron (for instance, between input neurons and hidden neuron z_1 in Fig. 10.1) is

$$\Delta \mathbf{w}_h^{t+1} = w_{0h} \eta \mathbf{X}^T \mathbf{Q}_h \mathbf{e} + \alpha \Delta \mathbf{w}_h^t \tag{10.23}$$

10.3.3 Procedure

During learning, the above equations will be used iteratively until some criteria are satisfied. The learning procedure will be

♦ Step 1, Initialization: assigning random values to all network parameters;
♦ Step 2, Estimation: estimate model outputs and errors by feeding input vectors;
♦ Step 3, Update: update all the model parameters using the above update rules;
♦ Step 4, Check: check if the desired criteria are satisfied, if so stop, otherwise go to Step 2.

There are commonly three stop criteria for use. They are the maximum learning cycle, the error threshold and the stability. A learning process will be terminated if the learning cycle has exceeded the maximum learning cycle. In some situations, if the training error has already been below the desired error threshold, a learning process will also be halted. For some complicated learning problems, we may not be interested in reaching the maximum learning cycle and may not be able to set a proper error threshold. In this case, we can check if the change of weights is small enough. There are two possible reasons for there being nearly no change in weights. First, a model has been well-trained whilst the desired error threshold is too small and the maximum learning cycle is too long. Second, an inappropriate setting of the learning parameters (the learning rate, the momentum factor, and the number of hidden neurons) leads to bad learning. If this happens, a learning process must be stopped manually to reset the learning parameters. In most cases, a large learning rate may end up with a pre-matured learning process where the change of weights will diminish much earlier than it should.

10.4 Applications to bioinformatics

We discuss some applications of neural networks to bioinformatics projects in this section.

10.4.1 *Bio-chemical data analysis*

Quantitative structure-activity relationship (QSAR) models are a class of bio-chemical models and are normally involved with binary input variables for chemical properties with a very large dimensionality. The use of neural networks is normally for relational study or dimensionality reduction. Each input vector in these applications therefore represents a binary vector, i.e. $\mathbf{x} \in \{0,1\}^D$. Each input vector is associated with a target value. In order to find the mapping function relating the chemical properties with the compound property, classification analysis approaches can be used. Neural networks can be used in these tasks for nonlinear modelling. For instance, a recent study using neural networks looked at the inhibition function of mutant PfDHFR [297]. In microbiological research, Bacillus species identification is not an easy task. The application of neural networks on 1071 fatty acid profiles has proved to be a powerful tool for this identification [298]. The neural networks have also been applied to the study of the relationship between compound chemical structures and human estrogen receptor (α and β) binding affinity, where the inputs are the molecular descriptors calculated from docking methods [299]. Heparanase inhibitors' activity was also predicted using neural networks based on QSAR data [300].

10.4.2 *Gene expression data analysis*

Gene expression data have been widely studied for understanding how genes respond to external environmental cues. Gene expression data are normally numerical inputs, also of a large dimensionality, but consisting of a few number of samples. In this case, data significance is a very serious problem in applying neural networks for data analysis. In recent studies, gene expression data have been used for disease diagnosis. In these applications, the expressions of genes are commonly sitting in a high dimensional space ($\mathbf{x} \in \mathfrak{R}^D$, where D is the number of genes and $\mathbf{x}$ is a vector of the expression values for D genes). Each expression vector ($\mathbf{x}$) has an associated target value, declaring the corresponding sample disease-free or not. It can be seen that this is then a classification problem. If the relationship between expression vector and target is nonlinear, neural network is one of the candidates for model construction and prediction. For instance, neural networks were used for the investigation of the distinguishing power of childhood acute lymphoblastic leukaemia (ALL) diagnostic bone marrow [301], and for influenza identification based on microarray data [302]. Neural networks have also been used for gene network re-construction [303] and for cancer-related regulatory modelling [304].

10.4.3 *Protein structure data analysis*

Protein structures are always an important subject for studying how proteins are interacting with each other, forming complexes for cellular signalling in response to environmental cues. Wagner et. al. applied neural networks to the function prediction of inhibitory activity of serotonin and NF-kappaB [305]. It was found that the relationship between structure and activity is essential to cellular signalling for the inhibitory function of serotonin and NF-kappaB. In a study involving the detection of drug-induced idiosyncratic liver toxicity using QSAR data, it was reported that a neural network model was able to achieve 84% accuracy [306].

10.4.4 *Bio-marker identification*

In bioinformatics research, the identification of bio-markers has a great importance in bio-medical applications. The major purpose in these applications is to identify the most important identities which can be genes, compounds, chemicals, proteins or metabolites for predictive usages. This means that we need to combine classification analysis approaches with feature selection approaches to identify a minimum subset of input variables which can achieve maximum discrimination capability between disease and disease-free samples. For instance, surface-enhanced laser desorption/ionization time-of-flight mass spectrometry was used to detect proteomic patterns in the serum of women with endometriosis [307]. Neural networks have been used for detecting early stage epithelial ovarian cancer using multiple serum markers from four institutes [308].

10.5 A case study on *Burkholderia pseudomallei* gene expression data

We use the reduced data set with ten top genes discovered in Chapter 8 for this demonstration. The ten top genes are BPSL2697, BPSL2522, BPSS1512, BPSS0477, BPSL2096, BPSS1525, BPSL2520, BPSS0476, BPSS1532, and BPSS1532.1. Hidden neurons are varied from two to 20. Five-fold cross-validation is used. First, the AUR and total prediction accuracy are used to select the best model (the highest performance measurements being either AUR or total accuracy). The left panel of Fig. 10.6 shows the ranking result using AUR. It shows that the model employing 17 hidden neurons demonstrates the best model robustness. However, when we treat the total prediction accuracy as the priority we find that the model employing two hidden neurons is the best one, as shown in the right panel of Fig. 10.6.

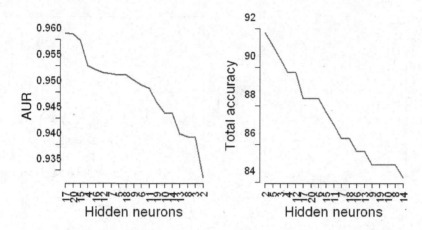

Fig. 10.6. Ordered AUR (left panel) and total accuracy (right panel) of the MLP models constructed for the *Burkholderia pseudomallei* gene data. The horizontal axes represent the number of hidden neurons. The vertical axes represent two performance measurements.

The detailed performance measurements of the model employing 17 hidden neurons is shown in Fig. 10.7, where we can see that some model predictions fall in the area between two clusters (left cluster of non-infected patients and the right cluster for the infected patients) shown in the left panel of Fig. 10.7 using density analysis. Checking the performance measurements, we find that although the specificity is 90%, the sensitivity is 93% as shown in the right panel of Fig. 10.7. This implies that the model predictions falling in the middle in the left panel are largely misclassification of the infected patients.

We then examine the model with two hidden neurons. The result is shown in Fig. 10.8. It can be seen that there are very few model predictions falling in the middle area between two clusters using density analysis (the left panel of Fig. 10.8). Compared with the density analysis in Fig. 10.7, we can see that the left cluster in the left panel of Fig. 10.8 has a very small variation. The specificity and the sensitivity are 93% and 88% shown in the right panel of Fig. 10.8.

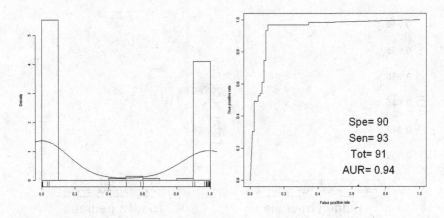

Fig. 10.7. The histogram of model outputs (left panel) and ROC curves (right panel) for the model using 17 hidden neurons.

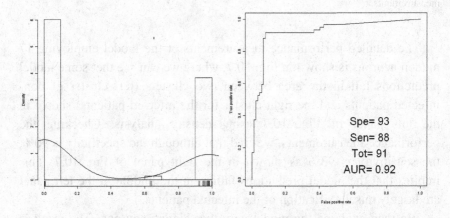

Fig. 10.8. The histogram of model outputs (left panel) and ROC curves (right panel) for the model using two hidden neurons. In the ROC curve, the horizontal axes represent 1 – specificity and the vertical axes represent sensitivity.

Figure 10.9 shows a further analysis of model selection using AIC and BIC. They are all consistent with the model selection result using the total prediction accuracy. We can then be confident in saying that the model employing two hidden neurons is the best model for this data.

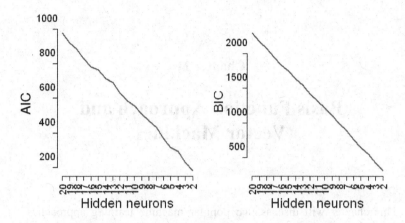

Fig. 10.9. The demonstration of model selection using AIC and BIC for the MLP models built for the *Burkholderia Pseudomallei* gene expression data.

Summary

This chapter has discussed the theory of multi-layer perceptron (MLP) and its application to bioinformatics. Through this discussion, we can see that 1) MLP is a nonlinear algorithm; 2) MLP can handle any function approximation problems; 3) MLP is an easy tool for modelling biological data with good performance. However, MLP has been criticised as being a *black-box* algorithm because it is difficult to know what the model parameters mean. To overcome this limitation, various researchers have been working on analysing the MLP weights. For details of this the reader may refer to Bishop's book [159].

Chapter 11

Basis Function Approach and Vector Machines

This chapter will discuss two popular machine learning approaches. They are basis function neural networks and vector machine algorithms. These two approaches have a similar background in machine learning, i.e. being non-parametric approaches for model construction. However, they have a fundamental difference in that the former will keep all the training data but the latter will use part of the training data for the inference process. This fundamental difference has given vector machine models better generalisation capability for unseen data. Their applications to bioinformatics are discussed as well in this chapter.

11.1 Introduction

In Chapter 3, the non-parametric kernel approach has been discussed where the density of a data set is estimated by

$$p(\mathbf{x}) = \frac{1}{N} \sum_{n=1}^{N} p(\mathbf{x} \mid \mathbf{x}_n, \vartheta) \qquad (11.1)$$

where $\mathbf{x} \in \Re^d$ is an input vector, $\mathbf{x}_n \in \Re^d$ is the nth training data, ϑ is a smooth parameter, $p(\mathbf{x} \mid \mathbf{x}_n, \vartheta)$ is a kernel function measuring the similarity between $\mathbf{x}$ and $\mathbf{x}_n$ using a pre-defined normal density function which has a smooth parameter ϑ, and N is the number of training data points. We can generalise equation (11.1) to the following format

154

$$p(\mathbf{x}\,|\,k) = \sum_{n=1}^{N_k} w_n^k p(\mathbf{x}\,|\,\mathbf{x}_n^k, \vartheta_n^k), \forall\, k \in [1, K] \qquad (11.2)$$

where k is the kth class, N_k is the number of training vectors in the kth class, K is the total number of classes in a data set, $\mathbf{x}_n^k$ is an input vector with a label of class k, ϑ_n^k is the smoothing parameter of the nth kernel function of the kth class, and $w_n^k \in \Re$ is the coefficient of the nth kernel of the kth class. $p(\mathbf{x}\,|\,k)$ is used to measure how likely $\mathbf{x}$ is to be generated by the kth class. An illustration is shown in Fig. 11.1 where a univariate data set with two classes is modelled. At a point where $\mathbf{x} = 2$, there are two probabilities (densities), being $p(x = 2\,|\,1)$ and $p(x = 2\,|\,2)$. Each of these two probabilities can be calculated in various ways. Using MLP mentioned in Chapter 10 is one method. Basis function neural network and vector machines which are the implementation of kernel approach are another two ways we discuss in this chapter.

The basic principle of basis function neural networks and vector machines is to estimate w_n^ks to construct a predictive model for a given data set. Because they are based on different statistical assumptions and use different learning mechanisms, the estimated w_n^ks will not be identical for the same data, hence leading to different performances for the two types of algorithm/approach. One important difference between the former and latter is that a basis function neural network does not generate a parsimonious model directly while vector machines aim to obtain a parsimonious model directly during a learning process. When a large data set is encountered, basis function neural networks need to employ a post-analysis or embed a procedure such as feature selection (which will be discussed in detail in Chapter 13) to simplify a model structure.

In this chapter, we discuss two basis function neural networks - the radial basis function neural network and the bio-basis function neural network - and two vector machines - the support vector machine and the relevant vector machine.

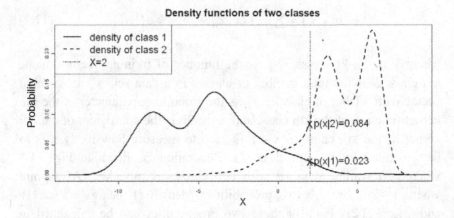

Fig. 11.1. An illustration of the density functions of two classes in a univariate data set. The horizontal axis represents the univariate X and the vertical axis represents the densities of two classes, one being marked by a solid line and the other being a broken line. The vertical line at X=2 indicates a prediction.

11.2 Radial-basis function neural network (RBFNN)

Let's denote a data set as $\mathcal{D} = \{ \mathbf{x}_n \in \mathfrak{R}^d \}_{n=1}^N$. Based on the same assumption used in the kernel density estimation approach that each input vector is randomly sampled from an infinite number of input vectors surrounding it with a Gaussian distribution, radial-basis function uses a Gaussian-like kernel function as below

$$\phi(\mathbf{x} \mid \mathbf{x}_n, \beta_n) = \exp(-\beta_n \| \mathbf{x} - \mathbf{x}_n \|^2) \qquad (11.3)$$

Figure 11.2 shows the radial-basis function with different β values. As the smoothing parameter increases, the radial-basis function becomes more sharply peaked.

The RBFNN model output is defined as

$$y = w_0 + \sum_{n=1}^N w_n \phi(\mathbf{x} \mid \mathbf{x}_n, \beta) \qquad (11.4)$$

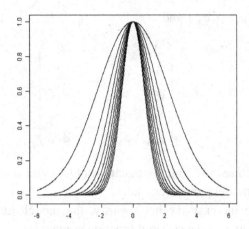

Fig. 11.2. An illustration of radial-basis functions with differential smoothing parameter values. The horizontal axis represents the variable X, the centre is zero. The vertical axis represents the output of the radial-basis functions.

Here a uniform smoothing parameter is used. There are two treatments to consider in using the model, one is for the regression mode and the other is for the classification mode. In regression mode, it is assumed that the target t is sampled with added noise of Gaussian distribution distributed in Gaussian

$$t_i = y_i + e_i \tag{11.5}$$

where e_i is the error (added noise), t_i is the ith target and y_i is its corresponding model output. The objective function using the least square error function with a regularisation term is defined as below

$$O = \frac{1}{2}\left(\sum_{i=1}^{N} \varepsilon_i^2 + \lambda \sum_{i=1}^{N} w_i^2\right) \tag{11.6}$$

where λ is a Lagrange constant. Letting the derivative of the objective function with respect to $\mathbf{w}$ be zero leads to

$$\mathbf{w} = (\mathbf{\Phi}^{\mathrm{T}}\mathbf{\Phi} + \lambda\,\mathbf{I})^{-1}\mathbf{\Phi}^{\mathrm{T}}\mathbf{t} \tag{11.7}$$

where $\mathbf{e} = (e_1, e_2, \cdots, e_N)$ and

$$\mathbf{\Phi} = \begin{pmatrix} 1 & 1 & \cdots & 1 \\ \phi_{11} & \phi_{12} & \cdots & \phi_{1N} \\ \phi_{21} & \phi_{22} & \cdots & \phi_{2N} \\ \vdots & \vdots & \vdots & \vdots \\ \phi_{N1} & \phi_{N2} & \cdots & \phi_{NN} \end{pmatrix} \tag{11.8}$$

With a pre-defined λ and β, equation (11.7) can directly lead to the estimation of model parameters, i.e. $\mathbf{w}$.

In order to increase the nonlinearity, the model output can also be defined as below using a sigmoid function conversion

$$y = \rho \left[w_0 + \sum_{n=1}^{N} w_n \phi(\mathbf{x} \,|\, \mathbf{x}_n, \beta) \right] \tag{11.9}$$

where $\rho(x)$ is a sigmoid function defined as

$$\rho(x) = \frac{1}{1 + \exp(-x)} \tag{11.10}$$

The derivative function of the objective function with respect to w_n is

$$\nabla \mathcal{L}(w_n) = -\sum_{m=1}^{M} e_m y_m (1 - y_m) \phi(\mathbf{x}_m' \,|\, \mathbf{x}_n, \beta) + \lambda \, w_n \tag{11.11}$$

The vector-matrix format of the derivative is shown as below

$$\nabla \mathcal{L}(\mathbf{w}) = -\mathbf{\Phi}^T \mathbf{\Lambda} \mathbf{e} + \lambda \, \mathbf{w} \tag{11.12}$$

where $\mathbf{\Lambda} = \mathrm{diag}\{ y_m (1 - y_m) \}$ is called an entropy matrix. Letting this derivative be zero leads to

$$\mathbf{w} = \frac{1}{\lambda} \mathbf{\Phi}^T \mathbf{\Lambda} \, \mathbf{e} = \frac{1}{\lambda} \mathbf{\Phi}^T \mathbf{\Lambda} \, (\mathbf{t} - \mathbf{y}) \tag{11.13}$$

Note that **y** is a function of **w**. The above equation cannot be used for estimating model parameters directly. Two procedures can be used for estimating model parameters. One is called the expectation-maximisation (EM) algorithm [157-159] and the other is called the stochastic algorithm. With the EM algorithm, we assign random values to **w** at first. Based on the current value for **w**, **y** values can be calculated using equation (11.9). This then leads to the update of **w** using equation (11.13). After a few iterations, **w** can be estimated. With the stochastic algorithm, we use the gradient descent approach which is defined as

$$\Delta \mathbf{w} = -\eta \, \nabla \, f(\mathbf{x}, \mathbf{w}) \qquad (11.14)$$

where $\eta \in (0,1)$ is called a learning rate. The update of **w** is also iterative. In each iteration the update of **w** is defined as

$$\mathbf{w}^{t+1} = (1 - \eta \lambda) \, \mathbf{w}^t + \eta \, \mathbf{\Phi}^T \mathbf{\Lambda} \, \mathbf{e} \qquad (11.15)$$

For both the EM and stochastic algorithms, the update continues until the maximum learning cycle is approached or the error is less than the pre-defined error threshold. Stopping a learning process when model parameters are in the stable status, i.e. no change in subsequent iterations, is also a commonly used approach.

In the classification mode, a different objective function is commonly used. For instance, the cross-entropy function is used for two-class classification problems. It is defined as

$$O = -\sum_{i=1}^{N} \left[t_i \log y_i + (1 - t_i) \log(1 - y_i) \right] + \frac{1}{2} \lambda \sum_{i=1}^{N} w_i^2 \qquad (11.16)$$

The derivative of this objective function with respect to **w** is

$$\nabla \mathcal{L}(\mathbf{w}) = -\mathbf{\Phi}^T \mathbf{e} + \lambda \, \mathbf{w} \qquad (11.17)$$

Because $\mathbf{e} = \mathbf{t} - \mathbf{y}$, where **y** is a function of **w**, this model cannot be solved explicitly. Both the EM algorithm and the stochastic algorithm can be used to estimate model parameters based on the above equation.

RBFNN has been intensively used in analysing biological data. For instance, it has been used to estimate the kinetic parameters of a dynamic biological system [309], and for gene data analysis [310-314].

However, there is a pitfall in using RBFNN for modelling molecular sequence data. As mentioned above, it is assumed that there is an infinite number of input vectors surrounding each training input vector. This means that if a biological data set is represented using a discrete approach, RBFNN is not applicable. For instance, if four nucleic acids in a DNA sequence are represented by 1, 2, 3, and 4 [315], there is certainly no other data surrounding each training input vector. For a data set of 2-mer nucleic acids, the data is sparsely distributed in a two-dimensional space where the variance of each circle is zero. This means that the smoothing parameter for each kernel is an infinite value. Any finite smoothing parameter cannot appropriately model the real data distribution.

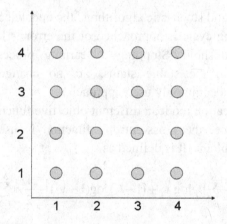

Fig. 11.3. An illustration of the discrete representation of biological data. Each circle represents a possible 2-mer peptide. Two axes represent the first and the second residue in a 2-mer peptide.

We now apply RBFNN to the *Burkholderia pseudomallei* gene expression data. Data are pre-processed as usual, i.e. logarithm is applied to remove the skew of the data. The data are then divided into five folds for cross-validation modelling. The smoothing parameter varies from

0.0005 to 0.01 with a step of 0.0005. Model performance is measured using the testing data set. Figure 11.4 shows the ranking of the models in terms of AUR (left panel) and the total prediction accuracy (right panel). Using AUR to rank models, the model with the smoothing parameter as 0.01 outperforms the others. Using the total prediction accuracy to rank the models, the model with the smoothing parameter as 0.01 is also the best.

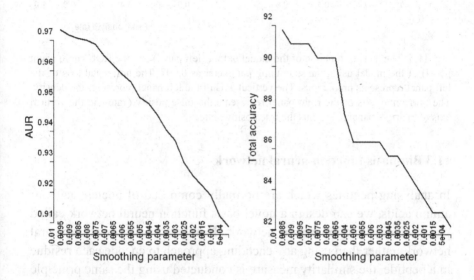

Fig. 11.4. The ranked RBFNN models built for the *Burkholderia pseudomallei* gene expression data using AUR (left panel) and the total prediction accuracy (right panel). The horizontal axes indicate the varying smoothing parameter while the vertical axes represent the performance measurements.

Details of the model using the smoothing parameter as 0.01 are shown in Fig. 11.5, where the left panel demonstrates the density of model outputs and the right panel shows the ROC curve as well as four measurements. The density of the model outputs clearly groups two classes of data (patients) together. This is why this model has the highest AUR, i.e. being the most robust model among all.

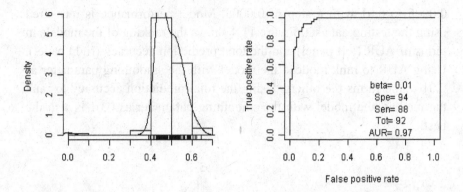

Fig. 11.5. The density function of the model output (left panel) and the ROC curve (right panel) of the model using the smoothing parameter as 0.007. The horizontal axis of the left panel represents predictions. The vertical axis of the left panel represents the density. The horizontal axis of the right panel represents the false positive rate and the vertical axis of the right panel represents the true positive rate.

11.3 Bio-basis function neural network

In analysing peptides which are normally composed of nucleic acids or amino acids, we can design a novel basis function neural network called the bio-basis function neural network. In the bio-basis function neural network, rather than using any encoding approach to encode each residue in a peptide, the similarity measure is conducted using the same principle as in sequence homology alignment.

When aligning two whole protein sequences, insertions and deletions are considered [10-14]. However, in handling short sequences or peptides which are normally less than 20 residues, insertions and deletions are normally not used. Using the homology alignment approach, two metrics can be used to score the similarity or distance between two peptides. They are the binary score such as the one used the Needleman-Wunsch algorithm [11] and the one used the Dayhoff algorithm as well as its variants [10, 15]. The Dayhoff score is also called a mutation matrix which is a 20 by 20 matrix for protein sequences, where each entry measures, for a particular pair of amino acids, the possibility that one amino acid is mutated to the other. It is therefore measuring the similarity between two amino acids and hence two sequences.

Before discussing the bio-basis function, we first discuss how to use binary score to handle the similarity between two peptides. For instance two nucleic peptides (AAC and AGC) can be expressed as 000100010010 and 000101000010. A binary similarity matrix is expressed as in Table 11.1. To quantify the similarity between two peptides, we can count the number of "1"s on the diagonal (expressed by italic number). This is similar to using dot product in some bioinformatics works [85, 86, 316]. Table 11.1 can actually be expressed by another simpler matrix shown in Table 11.2 where the similarity between two peptides is again the summation of the numbers on the diagonal.

Table 11.1. An illustration of a binary similarity matrix between two nucleic peptides. The first column represents AAC and the top row represents AGC. In the matrix, cells with empty entries indicate that the cells have zero entries.

	0	0	0	1	0	1	0	0	0	0	1	0
0												
0												
0												
1				*1*		1					1	
0												
0												
0												
1			1		1						1	
0												
0												
1			1		1						*1*	
0												

Table 11.2. An illustration of a simpler expression of binary similarity matrix between two nucleic peptides. The off-diagonal elements are ignored.

	A	G	C
A	1		
A		0	
C			1

Table 11.2 can be re-written as a new similarity matrix using a mutation matrix (PAM1 [5]) shown in Table 11.3. It can be seen that the summation of diagonal entries is slightly different from the result shown in Table 11.2.

Table 11.3. An illustration of a simpler expression of binary similarity matrix between two nucleic peptides. The off-diagonal elements are ignored.

	A	G	C
A	0.99		
A		0.00333	
C			0.99

Suppose two peptides are denoted by $\mathbf{s}_i \in \Theta^d$ and $\mathbf{s}_j \in \Theta^d$, where Θ is a set of nucleic acids or amino acids while d is the length of peptides. The bio-basis function is defined as below [317, 318]

$$z_{ij} = \phi(\mathbf{s}_i, \mathbf{s}_j) = \rho(-\beta\sigma(\mathbf{s}_i, \mathbf{s}_j)) \tag{11.18}$$

where ρ is a sigmoid function, β is a parameter measuring the sensitivity of a support peptide ($\mathbf{s}_j$) and

$$\sigma(\mathbf{s}_i, \mathbf{s}_j) = \sum_{r=1}^{d} M(s_{ir}, s_{jr}) \tag{11.19}$$

Here s_{ir} and s_{jr} are the rth residues of $\mathbf{s}_i$ and $\mathbf{s}_j$, respectively. From equation (11.14), we can see that if two peptides are identical, i.e. $\mathbf{s}_i \equiv \mathbf{s}_j$,

$$\lim_{\mathbf{s}_i \to \mathbf{s}_j} \phi(\mathbf{s}_i, \mathbf{s}_j) = 1, \forall \beta > 0 \tag{11.20}$$

However

$$\lim_{|\mathbf{s}_i - \mathbf{s}_j| \to \infty} \phi(\mathbf{s}_i, \mathbf{s}_j) = 0, \forall \beta > 0 \tag{11.21}$$

Note that we use the notation $|\mathbf{s}_i - \mathbf{s}_j| \to \infty$ to mean the distance between $\mathbf{s}_i$ and $\mathbf{s}_j$ is getting large. The model output also uses the sigmoid function as in equation (11.9). The negative log-likelihood function with added regularisation terms ($\lambda_w \sum_{n=1}^{N} w_n^2$ and $\lambda_\beta \sum_{n=1}^{N} \beta_n^2$) is

$$\mathcal{L} = -\sum_{m=1}^{M} [t_m \log y_m + (1 - t_m) \log(1 - y_m)]$$
$$+ \frac{1}{2} \lambda_w \sum_{n=1}^{N} w_n^2 + \frac{1}{2} \lambda_\beta \sum_{n=1}^{N} \beta_n^2 \qquad (11.22)$$

The derivative of the negative log-likelihood function with respect to β_n is

$$\nabla \mathcal{L}(\beta_n) = -\sum_{m=1}^{M} e_m w_n z_{mn} (1 - z_{mn}) \phi_{mn} + \lambda_\beta \beta_n \qquad (11.23)$$

Using the stochastic learning algorithm we have the update rule for β_n as defined below

$$\Delta \beta_n = -\eta \nabla \mathcal{L}(\beta_n) = \eta \left(\sum_{m=1}^{M} e_m w_n z_{mn} (1 - z_{mn}) \phi_{mn} - \lambda_\beta \beta_n \right) \qquad (11.24)$$

or

$$\beta^{t+1} = (1 - \eta \lambda_\beta) \beta + \eta \, \mathbf{Z}^T \mathbf{\Lambda} \mathbf{\Phi} \mathbf{w} \qquad (11.25)$$

The weight update rule is defined as

$$\mathbf{w}^{t+1} = (1 - \eta \lambda_w) \mathbf{w}^t + \eta \, \mathbf{Z}^T \mathbf{e} \qquad (11.26)$$

There are two modes in BBFNN, one being homogeneous and the other being heterogeneous. Using the homogeneous mode,

$$\beta_1 = \beta_2 = \cdots = \beta_N = \beta \qquad (11.27)$$

Using the same procedure discussed above for RBFNN, we can estimate the model parameters to build predictive models. The model built this way is called the bio-basis function neural network (BBFNN) which has been applied to various peptide classification tasks, for instance, the prediction of Trypsin cleavage sites [318], the prediction of HIV cleavage sites [317], the prediction of Hepatitis C virus protease cleavage sites [319], the prediction of the disorder segments in proteins [105, 320], the prediction of protein phosphorylation sites [319, 321], the prediction of the O-linkage sites in glycoproteins [322], the prediction of signal peptides [323], the prediction of factor Xa protease cleavage sites [324], the analysis of mutation patterns of HIV-1 drug resistance [325], the prediction of Caspase cleavage sites [326], the prediction of SARS-CoV protease cleavage sites [327] and T-cell epitope prediction [328].

Here we apply BBFNN to a peptide classification problem. The task is to predict HIV-I protease cleavage sites in a protein. HIV (human immuno-deficiency virus) is a retrovirus which causes AIDS (aquired immune deficiency syndrome) [329, 330]. HIV-I protease is an aspartic protease. an enzyme It plays an important role in the viral life-cycle. Each new infectious HIV viron is composed of mature protein components generated through cleaving a newly synthesised polyprotein at some specific sites in it using an HIV protease. These specific sites are called cleavage sites. Each cleavage site is a bond between two residues. The cleavage breaks down a polyprotein into functional components. If HIV proteases have been inhibited by a drug (the enzymes becoming ineffective), the HIV virons remain uninfectious [331, 332]. The inhibition through using drugs is then the major research focal point in fighting against the disease. In order to achieve this goal, it is important to design inhibitors to prevent the cleavage activities that produce new protease and reverse transcriptase. Studying how substrate specificity is related to cleavage activity is then critically important to the effect design of inhibitors. The HIV-1 protease data was published by Cai *et al.* [208]. The data contained 248 non-cleaved peptides and 114 cleaved peptides, each having eight residues. In using the homogeneous BBFNN we vary the sensitivity parameter from 0.1 to 10 with a gap of 0.1. Figure 11.6 shows the top 20 ordered performances. The left panel shows the ordered AUR which is maximised when the sensitivity parameter is 0.4. The right panel shows the ordered total prediction accuracy which is maximised when the sensitivity parameter is 0.5. The two are very close.

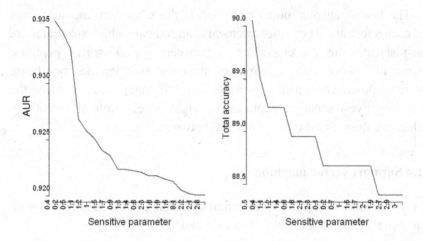

Fig. 11.6. The ordered top 20 performances using AUR (the left panel) and the total prediction accuracy (the right panel). The horizontal axes represent the 20 sensitivity parameters. The vertical axes represent the measured performances. 20-fold cross-validation is used for model evaluation.

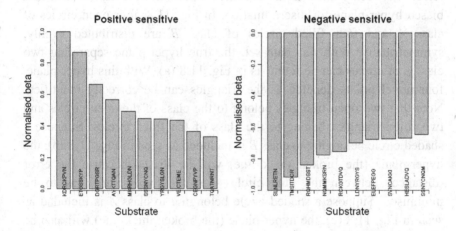

Fig. 11.7. The most positively and negatively sensitive peptides to the classification of cleaved and non-cleaved HIV substrates using the heterogeneous BBFNN model.

Using heterogeneous BBFNN for the same data can lead to similar performance with the specificity as 94%, the sensitivity as 80%, the total prediction accuracy as 90% and AUR as 0.94.

The heterogeneous model can provide the sensitivity measurements for each substrate. These measurements can indicate which substrates are most sensitive to the classification between two classes of peptides. Figure 11.7 shows two categories of most sensitive peptides, one being the most positively sensitive peptides (the left panel) and one being the most negatively sensitive peptides (the right panel). Note all sensitivity values are normalised into the interval between -1 and 1.

11.4 Support vector machine

A classification algorithm aims to find a mapping function between input features $\mathbf{x}$ and a class membership $t \in \{-1, 1\}$,

$$y = f(\mathbf{x}, \mathbf{w}) \tag{11.28}$$

where $\mathbf{w}$ is the parameter vector, $f(\mathbf{x}, \mathbf{w})$ is the mapping function and y is the model output. With other classification algorithms, the distance (error) between y and t is minimised to optimise $\mathbf{w}$. This can lead to a biased hyper-plane for discrimination. In Fig. 11.8, four open circles of class A and four filled circles of class B are distributed evenly, symmetrically. With this data set, the true hyper-plane separating two classes of circles can be found as in Fig. 11.8 (a). With this hyper-plane, four novel points denoted as the triangles can be correctly identified. Note that two open triangles belong to the class of the open circles and two filled triangles belong to the class of the filled circles. Suppose a shaded circle belonging to class B is included as seen in Fig. 11.8 (b), the hyper-plane (the broken thick line) will be biased because the error (distance) between the nine circles and the hyper-plane has to be minimised. Suppose a shaded circle belonging to class A is included as seen in Fig. 11.8 (c), the hyper-plane (the broken thick line) will also be biased. With these biased hyper-planes, the novel data denoted by the triangles could be misclassified.

In searching for the best hyper-plane, SVMs find a set of data points which are most difficult to classify. These data points are referred to as support vectors [333]. They are closest to the hyper-plane and are located on the boundaries of the margin between two classes. The advantage of

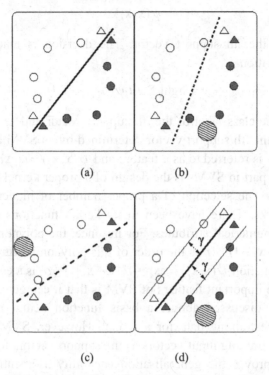

(a)   (b)

(c)   (d)

Fig. 11.8. (a) Hyper-plane formed using a conventional classification algorithm for the data with a balanced distribution. (b) and (c) Hyper-planes formed using a conventional classification algorithm for data without a balanced distribution. (d) Hyper-plane formed using SVMs for data without a balanced distribution. The open circles represent class *A*, the filled circles class *B*, and the shaded circle class *A* or *B*. The thick lines represent the correct hyper-plane for the discrimination and the broken thick lines the biased hyper-planes. The thin lines represent the margin boundaries. Gamma (γ) means the distance between hyper-plane and the boundary formed by the support vectors. The margin is 2γ.

using SVMs is that the hyper-plane is found through maximising this margin. Because of this, the SVM classifier is the most robust. Therefore it has the best generalisation ability. In Fig. 11.8 (d), two open circles on the upper boundary and two filled circles on the lower boundary are selected as support vectors. The use of these four circles can form the boundaries of the maximum margin between two classes. The trained SVM classifier is a linear combination of the similarity between an input and the support vectors. The similarity between an input and the support vectors is quantified by a kernel function defined as

$$\psi(\mathbf{x}, \mathbf{x}_i) \tag{11.29}$$

where $\mathbf{x}_i$ is the ith support vector. The decision is made using the following equation

$$y = \text{sign}\left(\Sigma \alpha_i t_i \psi(\mathbf{x}, \mathbf{x}_i) \right) \tag{11.30}$$

where t_i is the class label of the ith support vector and α_i the positive parameter of the ith support vector determined by an SVM algorithm. In SVMs, $\psi(\mathbf{x}_i)$ is referred to as a feature and $\psi(\mathbf{x}, \mathbf{x}_i) = \psi(\mathbf{x}) \cdot \psi(\mathbf{x}_i)$. The most difficult part in SVMs is the design of a proper kernel function that corresponds to the selection of a proper number of hidden neurons in neural networks. There have been many kernel functions designed for dealing with numerical attributes. For instance, the polynomial function $\psi(\mathbf{x}, \mathbf{x}_i) = (\mathbf{x} \cdot \mathbf{x}_i + 1)^P$ (p is the order of this polynomial function) or the radial basis function $\psi(\mathbf{x}, \mathbf{x}_i) = \exp(-\alpha | \mathbf{x} - \mathbf{x}_i |^2)$ (α is a constant).

One of the important features of SVM is that it can generate a sparse classifier. As discussed above, a basis function neural network will employ all the training data for a model. However, SVM will finally employ a few training input vectors as the support vectors for prediction. This first improves the generalisation capability as mentioned above. Second it can reduce model redundancy by removing unnecessary bases (kernels).

In application to whole protein sequences, the composition method has been the most popular method of analysis for many years. For instance, the composition method was used for the prediction of membrane protein types [334]. Dipeptides, gapped transitions (up to two gaps) and the occurrence of some motifs as additive numerical attributes were used to enhance the prediction of subcellular locations [335]. In the simulation it was shown that the inclusion of these additive numerical attributes did enhance the prediction accuracy. The same method has also been used in gene identification for functional RNAs in genomic sequences [336]. Instead of using transition composition to enhance the prediction performance, descriptors were also used, for instance, to predict multi-class protein folds [337]. SVMs also accurately discriminated cytoplasmic ribosomal protein genes from all other genes of a known function in *Saccharomyces cerevisiae*, *Escherichia coli* and

Mycobacterium tuberculosis using codon composition, a fusion of codon usage bias and amino acid composition sign [338].

There are two ways to generate profiles. First, a profile of a sequence can be generated by subjecting it to a homology alignment method like BLAST (Basic Local Alignment Search Tool) against a family of sequences in a database [10]. Second, a profile of a sequence can be generated using Hidden Markov Models (HMMs) [339]. For instance, HMMs were used to generate profiles based on positive sequences only and a Fisher kernel was designed for using SVMs to detect remote protein homologies [339]. The Fisher kernel was derived from the Fisher ratio, where the gradient vector of a sequence is computed with respect to the trained model. Each element of the gradient vector corresponds to a parameter of the HMMs. SVMs were trained on both positive and negative gradient vectors. Two methods (generating profiles using HMMs and homology alignment methods) have been compared for classifying G-protein coupled receptors [340]. The simulation showed that SVMs with HMMs profiles performed the best. The profile method was also used for the prediction of secondary structures [341].

Liao *et al.* used pair-wise homology alignment scores as features for training SVMs in protein homology detection [342]. An SVM classifier was then trained on these features. The work proved that this pair-wise-SVM performed better than Fisher-SVM [339]. SVMs were also used to classify proteins with remote homology into functional and structural families based on sequence homology using a newly designed string kernel function [343]. In that work, each feature is the occurrence of a specific K-mer (sub-sequence with K residues) in a sequence. Recently, SVMs were used to predict disordered regions in proteins, where a profile was formulated using PSI-BLAST (Protein Specific Iterated-BLAST) for each sequence against a non-redundant sequence database [344]. Moreover, SVMs were used to detect remote homology between protein sequences, which cannot be done sufficiently when using conventional methods like BLAST or FASTA (based on the idea of identifying short 'words' or k-tuples common to both sequences under comparison) [345].

In dealing with peptides, the orthogonal encoding method has been used for the analysis of molecular sequences using SVMs.

For instance, it was used for the prediction of translation initiation sites [346]. Interestingly, the work designed a novel kernel function which simply counted the number of nucleotides that coincide between two sequences. The kernel function was further improved based on the biological knowledge that local correlation information is important for translation initiation sites. It was also used for the classification of proteins with a selective kernel scaling method [347], for the classification of T-cell receptors [348], and for the prediction of protein-protein interactions [349].

Here we show how to apply SVM to the *Burkholderia pseudomallei* gene expression data in this section. Data pre-process is as usual. What we need to see here is how SVM can explore a few support *genes* (vectors) from all genes. The smoothing parameter varies from 0.001 to 0.1 with a gap of 0.001. The cost function is set at 1000. Figure 11.9 shows the ordered performance of models. The left panel shows the ranking result using AUR while the right panel shows the ranking result using the total prediction accuracy. Both show that model performance is optimised when the smoothing parameter is 0.002.

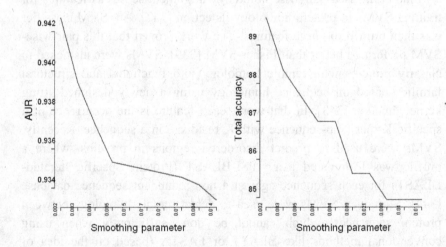

Fig. 11.9. The performance ranked SVM models for the *Burkholderia pseudomallei* gene expression data according to AUR (left panel) and the total prediction accuracy (right panel). In the ROC curve, the horizontal axes represent 1 – specificity and the vertical axes represent sensitivity.

Figure 11.10 shows how the support vectors are distributed. It can be seen that the closest neighbour of each support vector belongs to the class opposite to the class of the support vector. From this, it can be seen that SVM provides an excellent platform for data-mining biological data when exploring how individual biological components are contributing to the formation of a biological phenomenon.

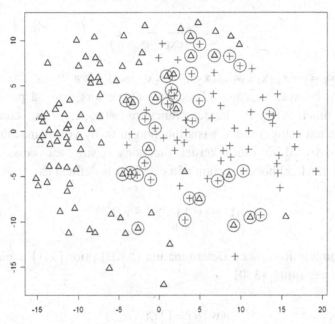

Fig. 11.10. Illustration of support vectors of the model using the smoothing parameter as 0.002. The triangle and the crosses represent two classes of patients. Those covered by circles indicate that they have been used as support vectors.

11.5 Relevance vector machine

In the above section, we can see that SVM estimates model parameters (**w**) through minimising the classification error and maximising the classification margin. The data distribution is not used in learning. RVM, which has the same kernel learning mechanism, instead, directly models the data distribution within the Bayesian framework [350]. Because of this, RVM can directly estimate the confidence of a prediction. The other important feature of RVM is that it is also a sparse classifier.

We denote by $\mathcal{D} = \{\mathbf{x}_n\}_{n=1}^{\ell}$ an input set and by $\{t_n\}_{n=1}^{\ell}$ a target set, where $\mathbf{x}_n \in \mathfrak{R}^d$ (d is the dimension) is an input vector and $t_n \in \mathsf{N}$ a target value. Note that $\mathfrak{R}$ is the set of real numbers and N is the set of integers. We use the sigmoid function to denote the relationship between an input vector and its prediction given a weight vector $\mathbf{w}$ [350]

$$y_n = \frac{1}{1 + \exp(-\boldsymbol{\varphi}_n \cdot \mathbf{w})} \tag{11.31}$$

where $\boldsymbol{\varphi}_n = (\phi(\mathbf{x}_n, \mathbf{x}_1), \phi(\mathbf{x}_n, \mathbf{x}_2), \cdots, \phi(\mathbf{x}_n, \mathbf{x}_\ell))^{\mathrm{T}}$ is a vector defining the similarity between $\mathbf{x}_n$ and all the training vectors using a pre-defined kernel function. The kernel function $\phi(\mathbf{x}_n, \mathbf{x}_m)$ is commonly implemented using a radial basis function in many vector machines and can be adapted to other kernel functions. Using the cross-entropy function, the likelihood function of a classifier is defined as

$$p(\mathbf{t} \mid \mathbf{w}) = \prod_{n=1}^{\ell} y_n^{t_n} (1 - y_n)^{1-t_n} \tag{11.32}$$

An Automatic Relevance Determination (ARD) prior [351] is placed to prevent over-fitting [350]

$$p(\mathbf{w} \mid \boldsymbol{\alpha}) = \prod_{n=1}^{\ell} \mathcal{G}(0, \alpha_n^{-1}) \tag{11.33}$$

where $\boldsymbol{\alpha} = (\alpha_0, \alpha_1, \alpha_2, \cdots, \alpha_\ell)^{\mathrm{T}}$. The posterior of the coefficients is defined as below

$$p(\mathbf{w} \mid \mathbf{t}, \boldsymbol{\alpha}) \propto |\boldsymbol{\Sigma}|^{-1/2} \exp\left\{-\frac{1}{2}(\mathbf{w} - \mathbf{u})^{\mathrm{T}} \boldsymbol{\Sigma}^{-1}(\mathbf{w} - \mathbf{u})\right\} \tag{11.34}$$

The mean vector and the covariance matrix of the posterior are

$$\mathbf{u} = \boldsymbol{\Sigma} \boldsymbol{\Phi}^{\mathrm{T}} \mathbf{B} \mathbf{t} \tag{11.35}$$

and

$$\Sigma = (\Phi^T B \Phi + A)^{-1} \tag{11.36}$$

where Φ is a squared input matrix $\Phi = \{\phi(x_i, x_j)\}_{1 \le i, j \le \ell}$, $t = (t_1, t_2, \cdots, t_\ell)^T$, $B = \text{diag}\{y_n(1 - y_n)\}$, and $A = \text{diag}\{\alpha_1, \alpha_2, \cdots, \alpha_\ell\}$. The marginal likelihood can be obtained through integrating out the coefficients [350]

$$p(t \mid \alpha) \propto |B^{-1} + \Phi A^{-1} \Phi^T|^{-1/2}$$
$$\exp\left\{-\frac{1}{2} t^T (B^{-1} + \Phi A^{-1} \Phi^T)^{-1} t\right\} \tag{11.37}$$

In learning, α can be estimated as follows

$$\alpha_n(\tau + 1) = \frac{1 - \alpha_n(\tau)\Sigma_{nn}(\tau)}{u_n^2(\tau)} \tag{11.38}$$

where τ is the iteration time and Σ_{nn} is the nth diagonal element in Σ. The weights can be updated using

$$\Delta w = -H^{-1} \nabla \log p(t, w \mid \alpha)|_{w_{MP}} \tag{11.39}$$

where $e = (e_1 = t_1 - y_1, e_2 = t_2 - y_2, \cdots, e_\ell = t_\ell - y_\ell)^T$, H is the Hessian matrix

$$H = \nabla \nabla \log p(t, w \mid \alpha)|_{w_{MP}} = -(\Phi^T B \Phi + A) \tag{11.40}$$

and

$$\nabla \log p(t, w \mid \alpha)|_{w_{MP}} = -(A w - \Phi^T e) \tag{11.41}$$

The above equation is a closed form where we have to use an inner loop for weight update.

Recently, RVM has drawn a lot of attention for analysing biological data. For instance, it has been used for predicting MHC-II binding affinity [352], for diagnosing cancers using gene expression profiles

[353], for inducing regulation transcription networks in Arabidopsis using gene expression data [354], and for detecting non-coding regions in genomes [355].

Summary

In this chapter we have discussed two classes of similar machine learning approaches, both employing the basic kernel approach. They are the basis function neural networks and vector machine algorithms. A basis function neural network model uses all the training input vectors for building a predictive model while a vector machine aims to find a sparse representation of all training input vectors by maintaining similar or improved prediction performance. The introduced radial-basis function neural network is used for handling numerical data. A note has been made in this chapter that the data used for building a radial-basis function neural network model must not be binary or sparsely distributed discrete data. The introduced bio-basis function neural network is used for handling sequence, particularly peptide data. The benefit of the bio-basis function neural network is that it avoids any tedious encoding process of amino acids or nucleic acids. Two vector machines are discussed in this chapter as well. They are the support vector machine and the relevance vector machine. A support vector machine model is generated by maximising the classification margin between two classes so that the generalisation capability of such a model can be maximised. The sparse model is therefore based on a subset of training input vectors. They are called support vectors. These support vectors are normally located on the decision boundaries of a classification margin and include some training vectors which are difficult to classify. The relevance vector machine, on the other hand, aims to find representative input vectors to avoid using all training input vectors in a model for prediction. The found relevance vectors are therefore those located in the centres of some clusters. The advantage of the relevance vector machine is that it is developed under the Bayesian framework, hence providing information of probabilistic interpretation of the predictions. Applications of these four algorithms to bioinformatics have also been discussed.

Chapter 12

Hidden Markov Model

This chapter discusses the hidden Markov model (HMM) which can be used to explore hidden series states for a sequence of observations. The basic principle and learning algorithm are discussed. The Markov model is discussed first as it provides the basis for understanding the hidden Markov model. Three basic tasks of HMM are discussed. They are the likelihood evaluation, decoding or prediction, and model parameter learning. Applications to bioinformatics are discussed as well.

12.1 Markov Model

An HMM is a statistical model where the statistical property of the transition probabilities between different observations and hidden state in a data set is modelled. The basis of HMM is the Markov process.

A Markov process describes a type of dynamic evolution systems where the relationship between random variables is mathematically defined. The model is named after the Russian mathematician Andrey Markov. In a Markov system, the likelihood of every possible random evolution process is evaluated using probability theory. A random evolution process is described for all possible random observations in a chain. The likelihood of each observation depends only on its previous observation in the same chain. Suppose we define a chain of T random variables as $(X_T, \cdots, X_t, \cdots, X_2, X_1)$ and have a chain of T observations denoted by $(x_T, \cdots, x_t, \cdots, x_2, x_1)$. The likelihood of observing x_t is defined as a conditional probability

$$P(X_t = x_t \mid \mathbf{X}_{t-1} = \mathbf{x}_{t-1}) = P(X_t = x_t \mid X_{t-1} = x_{t-1}) \qquad (12.1)$$

where $\mathbf{x}_{t-1} = (x_{t-1}, x_{t-2}, \cdots, x_2, x_1)$ and $\mathbf{X}_{t-1} = (\mathbf{X}_{t-1}, \mathbf{X}_{t-2}, \cdots, \mathbf{X}_2, \mathbf{X}_1)$. Based on the above equation, the likelihood of observing $(x_t, \cdots, x_2, x_1)$ is a product of all observation likelihood measurements using the product probability theory

$$
\begin{aligned}
P(\mathbf{X}_t = \mathbf{x}_t) &= P(x_t \mid x_{t-1}) P(x_{t-1} \mid x_{t-2}) \cdots P(x_2 \mid x_1) p(x_1) \\
&= P(x_1) \prod_{i=0}^{t-2} P(x_{t-i} \mid x_{t-i-1})
\end{aligned} \qquad (12.2)
$$

where $P(X_t = x_t \mid X_{t-1} = x_{t-1})$ simplified as $P(x_t \mid x_{t-1})$ and $P(x_1)$ is the probability of the first observation. This model is also called a first-order Markov model. Using the above equation the likelihood of observing a chain of five nucleic acids (TCGAA) shown in Fig. 12.1 is calculated as

$$P(A \mid A) P(A \mid G) P(G \mid C) P(C \mid T) P(T) \qquad (12.3)$$

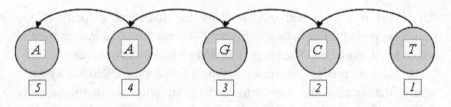

Fig. 12.1. A chain of five nucleic acids for demonstrating the Markov model. The numerical numbers represent the order of observing the nucleic acids. Five nucleic acids are represented by five letters.

If the transition probabilities are as specified in Fig. 12.2, the likelihood of the chain shown in Fig. 12.1 is $0.1 * 0.1 * 0.2 * 0.2 * 0.25 = 0.0001$.

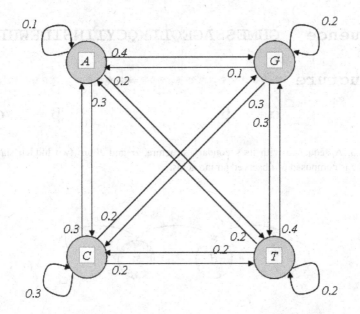

Fig. 12.2. The assumed transition probabilities between four nucleic acids. The summation of the transition probabilities from one nucleic acid to itself and to each of the other three nucleic acids is one. For instance, the transition probabilities from A to A, C, G, and T are 0.1, 0.3, 0.4, and 0.2 respectively.

12.2 Hidden Markov model

12.2.1 *General definition*

The Markov model described above only considers the probabilities of the transitions between observations. It is understood that various observations can result from some unknown hidden state. The observed events can be the phenomenon of some hidden genotypic information. For instance, we can consider the relationship between protein secondary structure and protein sequence. A sequence with corresponding secondary structures is shown in Fig. 12.3. The question is how we model the relationship between sequence residues and secondary structures leading to a model to identifying secondary structure based on the observed amino acid chain of a protein sequence.

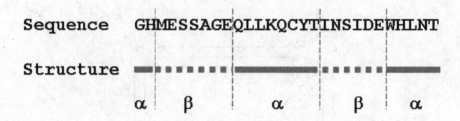

Fig. 12.3. A sequence with its secondary structure. α and β are two hidden states. A sequence is composed of observed amino acids.

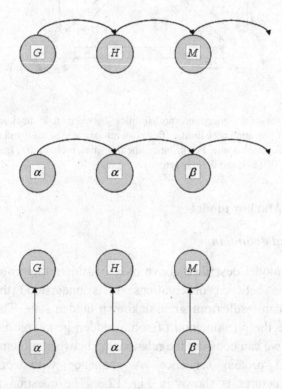

Fig. 12.4. A visualisation of three probabilities. The first row represents the Markov model. The second row represents the transition of hidden states. The last row represents the emission probabilities.

We can consider three sets of correlations. The correlation between each pair of residues measures how likely two residues are to become neighbours

$$\sigma(G,H), \sigma(H,M), \cdots \tag{12.4}$$

The correlation between each pair of states measures how likely two states are to be connected as neighbours

$$\sigma(\alpha, \alpha), \sigma(\alpha, \beta), \sigma(\beta, \beta), \sigma(\beta, \alpha) \tag{12.5}$$

The correlation between a residue and a state (secondary structure) measures how likely a residue and a state are to be aligned to the same position, for instance, the first residue G is aligned with the state E,

$$\sigma(\alpha, G), \sigma(\alpha, H), \sigma(\beta, M), \cdots \tag{12.6}$$

Figure 12.4 visualises these three correlations.

In HMM, the probability or the likelihood of the current observation not only depends on the previous observation, but also on the associated hidden state. The probability of one observation in a chain of observations is then defined as

$$P(X_{t+1} = x_{t+1} \mid X_t = x_t, S_t = s_t) \tag{12.7}$$

where S_t is the ith random variable of the hidden state in a chain of observations and $s_t \in \Theta$ is one of the hidden states associated with the ith observation. Θ is a finite set of hidden states. For instance, the third observation of the chain shown in Fig. 12.3 can be described as below

$$P(X_3 = M \mid X_2 = H, S_2 = \beta) \tag{12.8}$$

In HMM, such a probability is described as an emission model and the emission of the first three observations in the chain is shown in Fig. 12.5. In an emission model, each observation is emitted based on the transition probability from the previous observation to the current observation. It also depends on the transition probability from the previous hidden state to the current hidden state and the probability of emitting the current observation from the current state. For instance, the probability of emitting H in Fig. 12.3 is determined by the previous G, the transition from α to α, and the emission rate from α to H.

Based on the emission model shown in Fig. 12.5, a full HMM model for studying secondary structures of protein sequences is depicted in Fig. 12.6. In the diagram, we have three sets of probabilities to estimate. They are the emission probabilities, transition probabilities including self-transition probabilities, and terminal probabilities including start and end transition probabilities [158, 356].

The emission probabilities are defined by

$$\vartheta_S(X) = P(X \mid S) \tag{12.9}$$

where S indicates a state from a set of finite hidden states such as α and β shown in Fig. 12.3 and $P(X \mid S)$ is the probability of observing a phenomenon under a hidden state such as an amino acid in a sequence in an α secondary structure. The transition probabilities are defined as

$$\pi_{S_{t-1},S_t} = P(S_t \mid S_{t-1}) \tag{12.10}$$

where S_{t-1} is the (t-1)th state and S_t is the tth state. We use $\pi_{S,S}$ to denote self transition probabilities. The start transition probability is denoted by $\pi_{\phi,S}$, where ϕ means a terminal. The end transition probability is denoted by $\pi_{S,\phi}$.

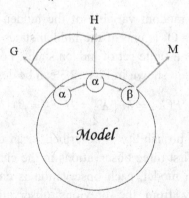

Fig. 12.5. An illustration of emitting the first three observations in a chain of observations shown in Fig. 12.3.

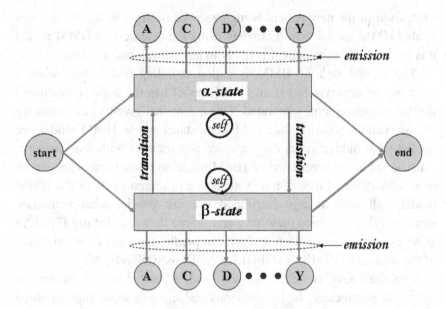

Fig. 12.6. An illustration of a diagram used in studying the relation between protein sequences and secondary structures.

12.2.2 *Handling HMM*

Having these probability definitions, we then discuss three tasks of HMM. These three tasks cover two theoretical and practical issues of HMM, parameter estimation and model interpretation. The three tasks are likelihood computing, decoding, and learning.

The first task is called evaluation and is to use the current model to interpret a sequence of observations, i.e. evaluating the likelihood that a sequence of observations is to be generated from a given HMM model. If we have a number of constructed HMM models each representing a specific biological function, we can evaluate from any of them the likelihood of a new sequence of observations being generated (emitted). For instance, we may have two HMM models constructed using gene expression data, one corresponding to disease-related patients, the other being related to disease-free patients. If the likelihood of the gene expression of the new patient being observed is larger using the disease-related HMM model than it is when using the disease-free HMM model it is predicted that the patient is likely to have developed a disease.

The second task of HMM is called decoding and is to decode a sequence of observations if an HMM model has the highest likelihood for that sequence being generated. This means that given a new sequence of observations without observed hidden states and an HMM model, we predict what hidden states the sequence is associated with. For instance, suppose we have constructed an HMM model to relate protein sequences to secondary structures. If a new sequence of amino acids fit the HMM model well with a large likelihood, we can predict what secondary structures this sequence may have and where they are. Taking Fig. 12.3 as an example, the question is if we can predict the secondary structures of the sequence GHMESSAGEQLLKQCYTINSIDEWHLNT.

The third task is called learning and is related to the estimation of model parameters. In the previous tasks, we assume that all three sets of probabilities are available. If these probabilities are not available, we need to estimate them to build an HMM model. In this situation, we are normally given a data set in which a number of sequences of observations and their hidden states are given. For instance, we may have collected a number of protein sequences each of which have experimentally verified secondary structures. Based on this data set, our job is to build an HMM model, i.e. to estimate three sets of probabilities. After this HMM model has been built, it can be used for the above two tasks.

12.2.3 *Evaluation*

If an HMM model has been built, i.e. its three sets of probabilities have been estimated, we can evaluate whether a new sequence is generated by this HMM model and the probability (likelihood) of this event. The evaluation is completed by a forward propagation of likelihood calculation using dynamic programming technique. The detail of dynamic programming is beyond the scope of this book, readers can refer to White DJ's textbook [357] and Bellman RE's textbook [358].

The use of dynamic programming for the evaluation is based on the Markov principle, i.e. each observation depends on its previous observation and hidden states as shown in equation (12.6). Suppose we have three possible observations A, B, and C and two hidden states α and β. The likelihood of the nth observation can be visualised in Fig. 12.7. The left panel shows the transition and emission probability calculations. However, having understood that we have already determined the nth observation and that the observations are controlled by hidden states, HMM involves simplifying the calculation as shown in the right panel of Fig. 12.7, where we only consider the state transition probabilities and the emission probabilities. The evaluation of the likelihood of the nth observation is then written as

$$P(X_t = x_t, S_t = s_t \mid X_{t-1} = x_{t-1}, S_{t-1} = s_{t-1})$$
$$= P(S_t = s_t \mid S_{t-1} = s_{t-1})P(X_t = x_t \mid X_{t-1} = x_{t-1}) \quad (12.11)$$

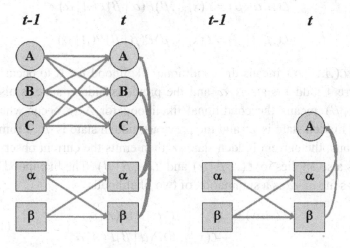

Fig. 12.7. An illustration of HMM for evaluating the likelihood of an observations fitting to an HMM model.

According to the product probability theory, the likelihood of two independent random events is the product of the likelihoods of these two events. We assume that the likelihood evaluated at the (n-1)th observation is independent from the likelihood evaluated at the nth observation. We can define the evaluation of the partial likelihood of the chain till the nth observation through a specific hidden state transition using the product of two likelihoods. One is the partial likelihood of the chain till the (t-1)th observation $L(x_{t-1}, s_{t-1})$ and the other is the calculation of the likelihood of the nth observation as defined in equation (12.11). The calculation of this single-path likelihood is then described as below

$$L(x_t, s_t) = L(x_{t-1}, s_{t-1})P(s_t \mid s_{t-1})P(x_t \mid s_t) \qquad (12.12)$$

For instance, there are four such calculations for the right panel in Fig. 12.7. They are

$$L(A, \alpha \mid \alpha) = L(x_{n-1}, \alpha)P(\alpha \mid \alpha)P(A \mid \alpha)$$

$$L(A, \beta \mid \alpha) = L(x_{n-1}, \alpha)P(\beta \mid \alpha)P(A \mid \beta)$$

$$L(A, \alpha \mid \beta) = L(x_{n-1}, \beta)P(\alpha \mid \beta)P(A \mid \alpha) \qquad (12.13)$$

$$L(A, \beta \mid \beta) = L(x_{n-1}, \beta)P(\beta \mid \beta)P(A \mid \beta)$$

where $L(A, \alpha \mid \alpha)$ means the conditional likelihood for A to occur when the current hidden state is α and the previous hidden state is also α. $L(A, \alpha \mid \beta)$ means the conditional likelihood for A to occur when the current hidden state is α and the previous hidden state is β. From both transitions, the current hidden state α then emits the current observation A. This also applies to $L(A, \beta \mid \alpha)$ and $L(A, \beta \mid \beta)$. The likelihood at the current state of α is a summation of two likelihoods

$$L(A, \alpha) = L(x_{n-1}, \alpha)P(\alpha \mid \alpha)P(A \mid \alpha)$$
$$+ L(x_{n-1}, \beta)P(\alpha \mid \beta)P(A \mid \alpha) \qquad (12.14)$$

This also applies to the current state β

$$L(A, \beta) = L(x_{n-1}, \alpha)P(\beta \mid \alpha)P(A \mid \beta)$$
$$+ L(x_{n-1}, \beta)P(\beta \mid \beta)P(A \mid \beta) \qquad (12.15)$$

We then have a likelihood calculation defined as below

$$L(x_t, s_t) = \sum_{k=1}^{\mathcal{K}} L(x_{t-1}, S_{t-1} = s_k) P(s_t \mid S_{t-1} = s_k) P(x_t \mid s_t) \quad (12.16)$$

where $\mathcal{K}$ is the number of states. When this evaluation for a sequence reaches the end terminal, the likelihood that an HMM model generates the sequence is calculated.

Suppose we have three probability matrices for the case shown in Fig. 12.3. The state transition matrix is shown in Table 12.1. The emission probability matrix for the first three residues in the sequence chain shown in Fig. 12.3 is shown in Table 12.2. The terminal transition probability matrix is shown in Table 12.3.

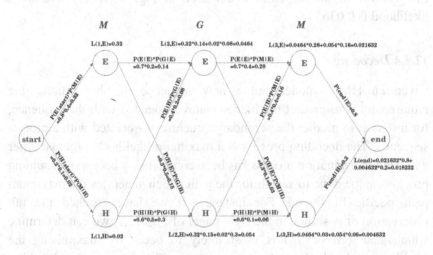

Fig. 12.8. An illustration of the likelihood evaluation of the first three residues in the chain shown in Fig. 12.3 with three probability matrices defined in Tables 12.1, 12.2, and 12.3.

Table 12.1. The state transition matrix for the case shown in Fig. 12.3.

	α	β
α	0.6	0.4
β	0.3	0.7

Table 12.2. The emission probability matrix for the first
three residues for the case shown in Fig. 12.3.

	G	H	M
α	0.5	0.4	0.1
β	0.2	0.4	0.4

Table 12.3. The terminal transition probability matrix for
the case shown in Fig. 12.3.

	α	β
Start	0.9	0.1
End	0.1	0.9

Based on these three matrices, Fig. 12.8 shows the likelihood of the
first residue in the sequence chain used in Fig. 12.3. The evaluated
likelihood is 0.036348.

12.2.4 *Decoding*

Given an HMM model and a new sequence of observations, the
requirement is to predict the hidden states associated with the sequence,
for instance, to predict the secondary structures associated with a protein
sequence. The decoding process is a maximum likelihood process. After
a likelihood evaluation process has been completed, a backward scanning
process can be done to search for the path which generates the maximum
path-specific likelihood. For instance, if we have reached the nth
observation of A shown in the right panel of Fig. 12.7, we can determine
which state (α or β) is most likely to occur by maximising the
likelihood calculated at these two states. The most basic algorithm
for decoding is the Viterbi algorithm [359]. Using the algorithm, we
can decode the network shown in Fig. 12.8. The predicted secondary
structures for the first three residues are seen in Fig. 12.9. The predicted
secondary structures are $\alpha\alpha\beta$ for the first three residues in the sequence.

HMM has also been well applied to analysing biological data. For
instance it has been used to identify orthologs in ESTs [360], for
predicting the occupancy of transcription factors in sequences [361],
for nucleic localisation signal prediction [362], for disease biomarker

identification [363], for predicting yeast gene functions [364] and for predicting cell wall sorting signals in gram-positive bacteria [365].

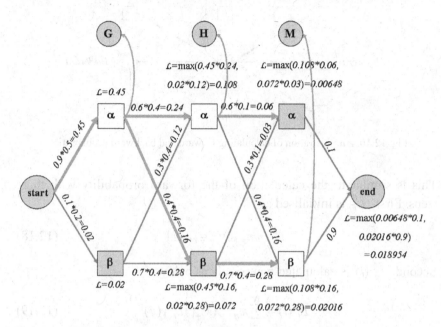

Fig. 12.9. An illustration of decoding an HMM for predicting the secondary structures of a protein sequence with observed amino acids.

12.2.5 *Learning*

Training an HMM means estimating the model parameters, i.e. the probabilities. The algorithm for solving this problem is called the Baum-Welch (BW) algorithm [366] which is a generalised EM algorithm [157]. With the BW algorithm, there are two parts of probabilities, one being the forward probability and the other being the backward probability. The forward probability is a probability of seeing the observations from the beginning to a node (marked as a filled circle in Fig. 12.10). The backward probability is calculated as below

$$\mathcal{L}_t^B(i) = P(x_{t+1}, x_{t+2}, \cdots, x_T \mid s_t = i, \lambda) \qquad (12.17)$$

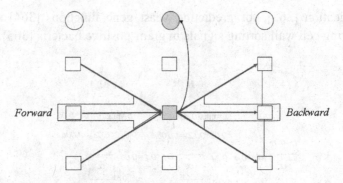

Fig. 12.10. An illustration of calculating forward and backward probabilities.

This is similar to the calculation of the forward probability with three steps. First, $\mathcal{L}_T^B$ is initialised as

$$\mathcal{L}_T^B = \pi_{i,\phi}, \forall\, i \in [1, N] \tag{12.18}$$

Second, $\mathcal{L}_t^B(i)$ is calculated recursively as below

$$\mathcal{L}_t^B(i) = \sum_{j=1}^N \pi_{ij} \vartheta_j(x_{t+1}) \mathcal{L}_{t+1}^B(j) \tag{12.19}$$

Finally, the calculation is terminated as

$$P(\mathbf{x} \mid \lambda) = \mathcal{L}_1^B(0) = \sum_{j=1}^N \pi_{\phi,j} \vartheta_j(x_1) \mathcal{L}_1^B(j) \tag{12.20}$$

After the calculation of both forward probabilities and backward probabilities, we can proceed to calculate the transition probabilities. The transition probability from the ith state to the jth state is defined as

$$\pi_{ij} = \frac{\text{expected number of transitions from } i \text{ to } j}{\text{expected number of transitions from } i} \tag{12.21}$$

The emission probabilities are calculated using

$$\vartheta_j(v_k) = \frac{\text{expected number of } times \text{ in } j \text{ and observing symbol } v_k}{\text{expected number of times in } j} \tag{12/22}$$

where v_k is one of the observed symbols. In the E-step, the partial forward and backward probabilities are calculated. In the M-step, the transition and emission probabilities are calculated.

12.3 HMM for sequence classification

HMM can be used for constructing predictive models for molecular sequences like other supervised machine learning algorithms. Details can be seen in Baldi's book [4] and Durbin's book [367]. HMMER [368, 369] is one of the most successful products and is composed of nine programs. Two main programs used for sequence classification are

A) "hmmbuild": builds a new profile HMM based on a data set in which sequences are aligned. The alignment of sequences can be done using various alignment algorithms.

B) "hmmpfam": aligns a set of sequences to the profile HMM generated by "hmmbuild" and outputs alignment scores and e-values.

In using HMMER for sequence (peptide) classification, an HMM profile is built using "hmmbuild" based on positive (functional) peptides [370]. After such an HMM profile has been generated, both positive and negative peptides are fed to the HMM profile using "hmmpfam" to obtain alignment scores. These e-values are then used to build two density functions for classification. Figure 12.11 shows the procedure of using HMM for sequence (peptide) classification, where steps 1, 2, 3, and 4 comprise a training process while step 5 is for testing.

We now use the HIV-1 protease cleavage data described in chapter 11 for demonstrating this process. The cleaved peptides (8-mers) are fed to the program called "hmmbuild" which generates an HMM profile. After the HMM profile has been generated, both negative (non-cleaved) and positive (cleaved) peptides are fed to the program called "hmmpfam" to generate two sets of alignment scores. Two Gaussian density functions are built. The Bayes rule is used to decide whether a novel peptide whose cleavage status is unknown is cleaved or non-cleaved. Five-fold cross-validation is used leading to the prediction performance as shown in Table 12.4.

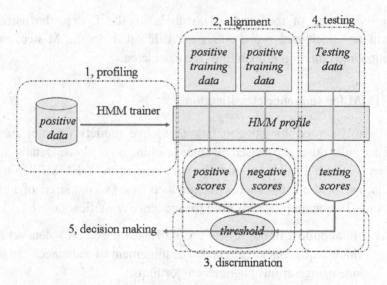

Fig. 12.11. A procedure of using HMM for sequence (peptide) classification. The dashed blocks represent five steps. The numbers (1, 2, 3, 4, and 5) represent the steps.

Table 12.4. The confusion matrix of applying HMMER to the HIV-1 protease cleavage data.

		Prediction		
		Negative	Positive	Percent
Actual	Negative	233	15	93.95%
	Positive	16	98	85.96%
				91.44%

The "hmmalign" program of HMMER is used to align peptides against an HMM profile that has been generated. Figure 12.12 shows the alignment of one of five positive alignments. The alignment for the positive peptides (left panel) against the built HMM profile shows a good convergence, i.e. nearly no insertion happens. However the alignment for the negative peptides (right panel) shows a large diversity with many insertions. This is why the alignment scores of the negative peptides are small by which we can see how HMM profiles can be used for classification. The basic principle is to learn the patterns hidden in the positive data with the belief that negative data serve as background, hence no pattern can emerge.

```
                        # STOCKHOLM 1.0
                        #=GF AU      HMMER 2.3.2

                        pep60          ....-KVFGRCEl...
                        pep61          .....--VFGRCEla..
                        pep65          ....CELAAAMK....
                        pep71          ....----MKRHgldn
                        pep73          .rhgLDNYR---....
                        pep79          ....YRGYSLGN....
                        pep91          ..ak----FESNfn..
                        pep100         ....QATNRNTD....
                        pep101         ....ATNRNTDG....
                        pep113         ....-GILQINSr...
                        pep116         ....QINSRWWC....
                        pep120         ....RWWCNDGR....
                        pep127         rtpgSRNL----....
                        pep135         .....--CNIPCSal..
                        pep146         ....DITASVNC....
                        pep149         ....ASVNCAKK....
                        pep150         ....SVNCAKKI....
                        pep155         ..kk----IVSDgn..
                        pep159         ....SDGNGMNA....
                        pep165         .....--NAWVAWrn..
                        pep167         ....-WVAWRNRc...
# STOCKHOLM 1.0          pep177         ....TDVQAWIR....
  #=GF AU      HMMER 2.3.2 pep184        ....-TAAAKFEr...
                        pep189         ....----FERQhmds
                        pep192         ....QHMDSSTS....
  pep3      SFNFPQIT    pep207         .cnq---MMKSR....
  pep5      ARVLAEAM    pep218         ...kDRCKPVN-....
  pep12     YEEFVQMM    pep220         ....-RCKPVNTf...
  pep16     AETFYVDK    pep223         ....PVNTFVHE....
  pep25     GDALLERN    pep225         .....--NTFVHEsl..
  pep29     AEAMSQVT    pep226         ....TFVHESLA....
  pep48     ELELAENR    pep236         ....QAVCSQKN....
  pep49     SKDLIAEI    pep246         ..ck---NGQTNc...
  pep53     PFAAAQQR    pep249         ....GQTNCYQS....
  pep57     AETFYTDG    pep250         ....QTNCYQSY....
  pep317    SQNYPIVE    pep253         .....---CYQSYstm.
  pep325    SFNYPQIT    pep255         .....-QSYSTMSi..
  pep332    SFNFPQII    pep256         ....SYSTMSIT....
  pep334    SQNYPNVQ    pep261         ....SITDCRET....
  pep337    SQNYPILQ    pep270         ....SSKYPNCA....
  pep340    SQCYPIVQ    pep275         ....NCAYKTTQ....
  pep346    ARVLFIAL    pep281         ....TQANKHII....
  pep351    ARVLFTAL    pep284         ....NKHIIVAC....
  pep352    ARNLFEAL    pep294         ....NPYVPVHF....
  pep353    ARNLFQAL    pep295         ....PYVPVHFD....
  pep356    ARVYPEAL    pep299         ....RQNYPIVQ....
  pep361    RQNYPIAL    pep300         ....SQKYPIVQ....
  #=GC RF   xxxxxxxx    pep306         ....SQNYDIVQ....
  //                    #=GC RF        ....xxxxxxxx....
                        //
```

Fig. 12.12. Alignments of peptides against the built HMM profile. The left panel shows the alignment for the positive peptides and the right panel shows the alignment for the negative peptides.

HMM has been widely used in sequence analysis, for instance it has been used for phosphorylation site prediction [370, 371], for predicting protein family [372], for modelling paramyxovirus hemagglutinin-neuraminidase proteins [373], for predicting the occupancy of transcriptional factors [361], for detecting recombination in 4-taxa DNA sequences [374], and for predicting genetic structure in eukaryotic DNAs [375].

Summary

This chapter has introduced the basic principle and the learning mechanism of hidden Markov models. Generally speaking, it learns the hidden states by which it is believed that observations are generated. HMM is a type of generative models which assumes that the observations are unorganised information of phenotypic and genotypic data while the relationship between them is hidden or unknown. Through learning, the relationship can be explored which can be used for pattern recognition. An example of HIV-1 protease cleavage peptide classification clearly shows the features of HMM.

Chapter 13

Feature Selection

Feature selection has long been studied in machine learning [376-379]. When analysing gene expression data and metabolite data it is common that a data set has a few samples with many genes or metabolites as variables. In order to focus on some highly differentially expressed genes or active metabolites for investigating biological insight, a feature selection process must be considered. The main task of feature selection is to reduce the number of features while maintaining or improving model predictive capability. This chapter discusses three types of feature selection strategy. The first is the built-in strategy. The second is the exhaustive strategy. The third is the heuristic strategy. The built-in strategy embeds a feature selection process in model construction. The typical algorithms include principal component analysis, the classification and regression tree as well as the random forest algorithm plus three other algorithms discussed in this chapter. The exhaustive strategy is to exhaust all possible models with different features and then select a model with the smallest number of features and best model performance. The evaluation is commonly based on AIC or BIC discussed in chapter 5. The heuristic strategy selects features step by step using an additive performance measure. This strategy includes forward and backward selection.

13.1 Built-in strategy

In the previous chapters, we have seen three relevant machine learning algorithms which can be used as well for feature selection. They are principal component analysis, the classification and regression tree algorithm as well as the random forest algorithm. They have a built-in process to remove irrelevant or unimportant variables (features) while maintaining features which are important for predictions. We are not going to discuss them in detail again in this chapter. Instead we introduce three other algorithms, i.e. the Lasso, the ridge regression, and the partial least square regression algorithms.

13.1.1 *Lasso regression*

The algorithm's full name is L1 constrained estimation 'Lasso'. It is a shrinkage and selection approach for linear regression. During learning, it minimises the sum of squared errors using a limit on the sum of the absolute values of the coefficients [380-382].

Given a set of independent variables $\{X_1, X_2, \cdots, X_R\}$ and a dependent variable Y, the Lasso model is defined as usual in a linear regression format

$$y = w_0 + \mathbf{x}^\mathrm{T}\mathbf{w} = w_0 + \sum_{i=1}^{R} w_i x_i \tag{13.1}$$

where x_i is the value of the ith variable X_i, w_0 is a bias, and w_i is the coefficient or weight for the ith variable. The error function which is minimised by Lasso is also similar to most regression models and is defined as below

$$\varepsilon = \sum_{n=1}^{N} (t_n - y_n)^2 \tag{13.2}$$

where N is the total number of input vectors, t_n is the nth target value, and y_n is the nth model output. However Lasso introduces a constraint as below

$$\sum_{i=1}^{R} |w_i| \le \tau \tag{13.3}$$

where $\tau > 0$ is the constraint constant. If τ is small, more coefficients are shrunk to zero. This means that unimportant variables are penalised while important variables are maintained in a model. This model is generally not analytically solvable and the quadratic programming approach [383] is employed. The algorithm called least angle regression [384] can also solve this problem.

In bioinformatics, Lasso is used to derive parsimonious or sparse regression models. For instance, it is used for building Cox proportional hazards models [385], for constructing gene networks through exploring mutual relationships between genes [386, 387], and for detecting causative genes of diseases [388].

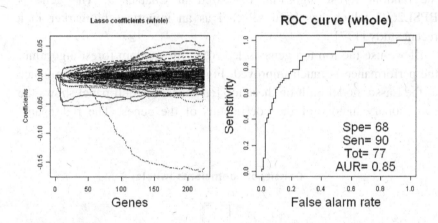

Fig. 13.1. The Lasso model applied to the *Burkholderia pseudomallei* gene expression data. The left panel shows the evolution of coefficients through learning iterations. The horizontal axis shows the learning iterations and the vertical axis shows the magnitudes of the coefficients. The right panel shows the ROC curves as well as the performance measures.

In applying Lasso to the *Burkholderia pseudomallei* gene expression data, the coefficients of most genes are penalised (reduced to zero). The left panel of Fig. 13.1 shows how coefficients are updated through learning iterations. It can be seen that for only a few genes the coefficients gradually evolve away from small values, getting larger and larger to reach a stable status. It is not surprising that the performance is not as good as some other machine learning models discussed in the previous chapters. This is due to the fact that Lasso is a purely linear regression model. It will not perform as well as nonlinear machine learning algorithms.

Figure 13.2 shows the density function of the coefficients of the Lasso model for the data. It can be seen that many coefficients are around the centre with a value of zero. On the left side, only two coefficients have values of less than -0.1. On the right side, only one coefficient has a value close to 0.1. Among ten top genes selected by the Lasso model, only one gene is consistent with the results obtained using the random forest algorithm discussed in Chapter 9. The gene is BPSL2697 which has been selected as an important biomarker in a recent study [177].

If we use the top ten genes selected by the random forest algorithm, the performance is much improved. Figure 13.3 shows the performance of the Lasso model built on these top ten genes. The left panel shows the evolutionary history of the coefficients of the genes. The right panel

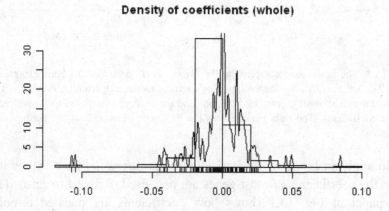

Fig. 13.2. The density function of the coefficients of the Lasso model built for the *Burkholderia pseudomallei* gene expression data. The horizontal axis represents the magnitudes of the coefficients while the vertical axis represents the density.

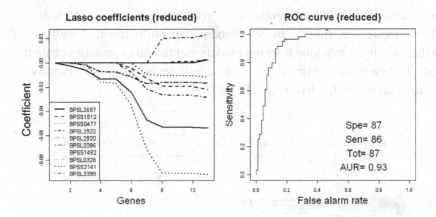

Fig. 13.3. The Lasso model built based on ten top genes selected by the random forest algorithm. The left panel shows the evolutionary history of coefficients update. The right panel shows the ROC curves as well as the performances. In the ROC curve, the horizontal axes represent 1 − specificity and the vertical axes represent sensitivity.

shows the ROC curves and the performance measurements. It can be seen that using the random forest algorithm to filter out noise variables can lead to better Lasso performance. This demonstrates a fundamental limitation of Lasso that the noise in data limits the selection of good variables.

13.1.2 *Ridge regression*

The ridge regression model was proposed in the 1970s for handling ill-posed linear algebraic equations [389, 390]. The weight decay [159, 391] used in neural learning since the 1980's is rooted from this. Rather than using L_1, the ridge regression approach uses L_2 constraint as below

$$\varepsilon = \sum_{n=1}^{N}(t_n - y_n)^2 + \lambda\sum_{i=1}^{R}w_i^2 \qquad (13.4)$$

Figure 13.4 shows the coefficients evolutionary history (left panel) and the ROC curves as well as prediction performances (right panel) based on whole *Burkholderia pseudomallei* gene expression data. As with the Lasso model, the ridge regression model does not perform as well as the other nonlinear machine learning models mentioned in previous chapters. Meanwhile it is as expected that the coefficients shrink consistently to zero as shown in the left panel of Fig. 13.4.

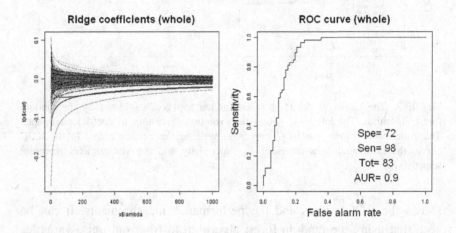

Fig. 13.4. The ridge regression model for the *Burkholderia pseudomallei* gene expression data. The left panel shows the coefficient evolutionary history, where the horizontal axis represents the learning iterations while the vertical axis represents the magnitudes of the coefficients. The right panel shows the ROC curve and the performance measurements. In the ROC curve, the horizontal axes represent 1 – specificity and the vertical axes represent sensitivity.

13.1.3 *Partial least square regression (PLS) algorithm*

PLS algorithm is different from Lasso and ridge regression in that it combines with principal component analysis (PCA) for selecting features. In chapter 4, we have discussed PCA which only maps the input matrix into an orthogonal space in which the variance in coordinates are ordered. If the first few (<3) principal components contain the majority of the information (variance) in data, they can be used for visualisation or can be used as features for further supervised learning. However, these principal components may not be very informative. For instance, if we run two different PCA simulations on the same data, one being based on the input data only and the other being based on both input data and the output data, we will see the difference. Two data sets are composed of two clusters in two dimensions, hence being two classes. We use $\mathbf{X}$ to denote the independent variable matrix and use $\mathbf{t}$ to denote the dependent variable vector. The first data set (the upper left panels in Figs. 13.5 and 13.6) has two clusters distributed in parallel to the X-axis. This means that the variable corresponding to the vertical axis is the only one contributing to perfect classification of two classes of data points. The second data set (lower left panels in Figs. 13.5 and 13.6) is generated by rotating the first data set. This means that both independent variables are contributing to the classification of two classes.

In the first simulation, we run PCA on $\mathbf{X}$. Figure 13.5 shows the results of both data sets. It can be seen from the right panels of the Fig. that for both sets of data, PCA gives similar eigen values (variances) to two independent variables. For the second, it makes sense because both independent variables are contributing to the classification. However, this is not true for the first data set.

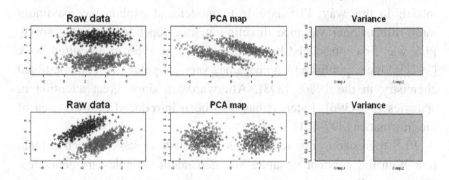

Fig. 13.5. An illustration of PCA on independent variables only for selecting features. Two rows are for two data sets. The left panels display the raw data distribution. The middle panels show the PCA maps. The right panels show the eigen values.

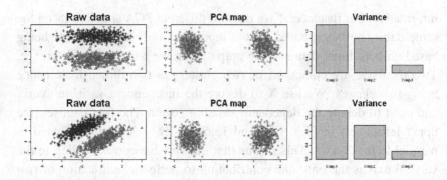

Fig. 13.6. An illustration of PCA on both independent and dependent variables for selecting features. Two rows are for two data sets. The left panels display the raw data distribution. The middle panels show the PCA maps. The right panels show the eigen values.

In the second simulation, we run PCA on $(\mathbf{X}, \mathbf{t})$. Figure 13.6 shows the results for both data sets. It can be seen from the right panels of the Fig. that there is a larger difference between the first and the second eigen values.

In PLS, principal components, relevant to the dependent variable, are found. This is why a PLS model is also referred to as a bilinear factor model. In this way, PLS is able to model and explain the maximum multi-dimensional variance direction in the dependent variable space. PLS was first introduced by Herman Word in 1966 in an edited book [392]. Since published, it became very popular in computational chemistry in the 1980s [393]. Afterwards, it drew great attention in statistics [394-396]. Later, it has also been introduced into the area of bioinformatics [397-401].

PLS regression aims to find a set of latent variables which explain how both independent variables and dependent variables are generated. Denote by $\mathbf{X}$ a matrix of N rows of the vectors for d independent variables and $\mathbf{y}$ a vector of N rows of the values for a dependent variable (it can also include more than one dependent variable). Both $\mathbf{X}$ and $\mathbf{y}$ are normalised with zero mean and one standard deviation. The latent variables or components are found step by step. The kth PLS component is obtained by estimating the corresponding weight vector $\mathbf{w}$ so that [402]

$$\mathbf{w}_k = \arg\max_{\mathbf{w}^T\mathbf{w}=1} \text{cov}(\mathbf{Xw}, \mathbf{y}) = \arg\max_{\mathbf{w}^T\mathbf{w}=1} (N-1)^{-1}\mathbf{w}^T\mathbf{X}^T\mathbf{y} \quad (13.5)$$

with the orthogonal constraints $\mathbf{w}_k^T \mathbf{S}\mathbf{w}_j = 0, \forall 1 \le j \le k$, where $\mathbf{S} = \mathbf{X}^T\mathbf{X}$.

Applying PLS to the *Burkholderia pseudomallei* gene expression data leads to the specificity as 87%, sensitivity as 93%, total prediction accuracy as 90%, and as AUR 0.94.

The density function of the coefficients is illustrated in Fig. 13.7, where we can see that only a few coefficients have large absolute magnitudes. The PLS has selected 5 genes with largest positive coefficients. They are BPSL0280, BPSS1993, BPSL0919, BPSL2298, and BPSL0665. The genes with most negative coefficients from the PLS model are BPSL2504, BPSL1631, BPSS2185, BPSS0796.1, BPSL3228, and BPSS1850.

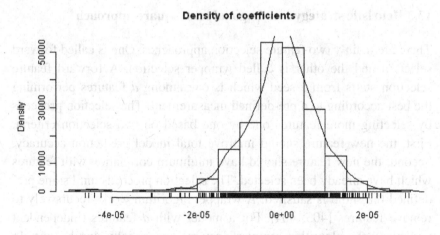

Fig. 13.7. A density function estimated for the coefficients in the PLS model for the *Burkholderia pseudomallei* gene expression data. The horizontal axis represents the magnitudes of the coefficients and the vertical axis indicates the density.

13.2 Exhaustive strategy

With the exhaustive strategy, all possible feature combinations must be exhausted. Each model with a specific combination of features is examined using AIC or BIC as discussed in the previous chapter [158, 159, 162]. The procedure is very straightforward, by preparing all the feature sets and constructing models based on these sets. After using AIC or BIC to evaluate them, the best model is selected for prediction. In this section, we evaluate this strategy by using the reduced *Burkholderia pseudomallei* gene expression data generated by the random forest algorithm in chapter 9. We exhaust all possible sets of three genes. MLP discussed in chapter 10 is used to model these data sets. The selected three genes which can yield the best performance are BPSL2697, BPSL 2522, and BPSL3398. The specificity is 91%, the sensitivity is 86%, the total prediction accuracy is 89% and AUR is 0.96. The ROC curve can be seen from the Figure.

It must be noted that the exhaustive strategy has limited usage in applications where the number of variables is large.

13.3 Heuristic strategy – orthogonal least square approach

There are mainly two feature selection approaches. One is called forward selection and the other is called wrapper selection. A forward feature selection starts from a seed which is one among d features performing the best according to a pre-defined measurement. The selection proceeds by selecting more features one by one based on two selection criteria. First, the new feature should improve total model prediction accuracy. Second, the new feature should have minimum correlation with features which have already been selected. The selection proceeds until some pre-defined threshold is satisfied. A wrapper algorithm works recursively to remove features [403, 404]. For a model with d features (independent variables), the algorithm removes features sequentially one by one. In each step, a feature is targeted if removing it can maximise the prediction accuracy.

The orthogonal least square (OLS) algorithm [405] is a forward selection procedure. At each step the incremental information content of a system is maximised. The feature matrix is denoted by $\mathbf{X} = (\mathbf{z}_1, \mathbf{z}_2, ..., \mathbf{z}_d)$. The OLS transforms the original variables ($\mathbf{z}_k$) to the orthogonal variables ($\mathbf{p}_k$) to reduce possible information redundancy. The feature matrix $\mathbf{X}$ is decomposed as

$$\mathbf{X} = \mathbf{PT} \tag{13.6}$$

where the triangular matrix $\mathbf{T}$ has 1's on the diagonal.

$$\mathbf{T} = \begin{pmatrix} 1 & t_{12} & t_{13} & \cdots & t_{1,d-1} & t_{1d} \\ 0 & 1 & t_{23} & \cdots & t_{2,d-1} & t_{2d} \\ 0 & 0 & 1 & \cdots & t_{3,d-1} & t_{3d} \\ \vdots & \vdots & \vdots & \vdots & \vdots & \vdots \\ 0 & 0 & 0 & \cdots & 1 & t_{d-1,d} \\ 0 & 0 & 0 & \cdots & 0 & 1 \end{pmatrix} \tag{13.7}$$

and the orthogonal matrix $\mathbf{P}$ is

$$\mathbf{P} = \begin{pmatrix} p_{11} & p_{12} & \cdots & p_{1d} \\ p_{21} & p_{22} & \cdots & p_{2d} \\ \vdots & \vdots & \vdots & \vdots \\ p_{N1} & p_{N2} & \cdots & p_{Nd} \end{pmatrix} = (\mathbf{p}_1, \mathbf{p}_2, ..., \mathbf{p}_d) \tag{13.8}$$

The orthogonal matrix satisfies

$$\mathbf{P}^T \mathbf{P} = \mathbf{H} \tag{13.9}$$

where $\mathbf{H}$ is diagonal whose elements h_{kk} :

$$h_{kk} = \mathbf{p}_k^T \mathbf{p}_k = \sum_{n=1}^{N} p_{nk}^2 \tag{13.10}$$

The space spanned by the set of orthogonal variables is the same space spanned by the set of original variables, and equation (13.6) can be rewritten as

$$\mathbf{y} = \mathbf{Xw} + \mathbf{e} = \mathbf{PTw} + \mathbf{e} = \mathbf{Pg} + \mathbf{e} \qquad (13.11)$$

Suppose $\mathbf{e} \sim N(\mathbf{0,1})$, the pseudo inverse method can be used to estimate $\mathbf{g}$ as below

$$\mathbf{g} = (\mathbf{P}^T\mathbf{P})^{-1}\mathbf{P}^T\mathbf{y} = \mathbf{H}^{-1}\mathbf{P}^T\mathbf{y} \qquad (13.12)$$

Because $\mathbf{H}$ is diagonal, its inverse matrix is shown as below

$$\mathbf{H}^{-1} = \begin{pmatrix} \dfrac{1}{h_{11}} & 0 & \cdots & 0 \\ 0 & \dfrac{1}{h_{22}} & \cdots & 0 \\ \vdots & \vdots & \vdots & \vdots \\ 0 & 0 & \cdots & \dfrac{1}{h_{dd}} \end{pmatrix}$$

$$= \begin{pmatrix} \dfrac{1}{\mathbf{p}_1^T\mathbf{p}_1} & 0 & \cdots & 0 \\ 0 & \dfrac{1}{\mathbf{p}_2^T\mathbf{p}_2} & \cdots & 0 \\ \vdots & \vdots & \vdots & \vdots \\ 0 & 0 & \cdots & \dfrac{1}{\mathbf{p}_d^T\mathbf{p}_d} \end{pmatrix} \qquad (13.13)$$

The element in $\mathbf{g}$ is then

$$g_k = \frac{\mathbf{p}_k^T\mathbf{y}}{\mathbf{p}_k^T\mathbf{p}_k} \qquad (13.14)$$

The quantities $\mathbf{g}$ and $\mathbf{w}$ satisfy the triangular system

$$\mathbf{Tw} = \mathbf{g} \qquad (13.15)$$

The Gram-Schmidt or the modified Gram-Schmidt methods [406-408] can be used for the selection, where the first variable is selected as the first orthogonal one $\mathbf{p}_1 = \mathbf{z}_1$. In the selection of the kth orthogonal variable, the elements in the kth column in $\mathbf{T}$ are estimated using the following equation

$$t_{ik} = \frac{\mathbf{p}_i^{\mathrm{T}} \mathbf{z}_k}{\mathbf{p}_i^{\mathrm{T}} \mathbf{p}_i}, \ 1 \leq i < k, \ k \in [2, d] \tag{13.16}$$

The kth orthogonal variable is then estimated as follows

$$\mathbf{p}_k = \mathbf{z}_k - \sum_{i=1}^{k-1} t_{ik} \mathbf{p}_i, \ k \in [2, d] \tag{13.17}$$

According to equation (13.15) we can estimate $\mathbf{w}$ as below

$$\mathbf{w} = (\mathbf{T}^{\mathrm{T}} \mathbf{T})^{-1} \mathbf{T}^{\mathrm{T}} \mathbf{g} \tag{13.18}$$

The elements in $\mathbf{w}$ exactly indicate which original variables are important in constructing the orthogonal variable space for modelling. OLS has recently been used for analysing gene expression data [409-411].

In using the OLS algorithm, we can terminate the iteration based on a pre-defined threshold. From equation (13.11), we can see that [405]

$$\mathbf{y}^{\mathrm{T}} \mathbf{y} = \mathbf{g}^{\mathrm{T}} \mathbf{P}^{\mathrm{T}} \mathbf{P} \mathbf{g} + \mathbf{e}^{\mathrm{T}} \mathbf{e} \tag{13.19}$$

From equation (13.9), we then have

$$\mathbf{y}^{\mathrm{T}} \mathbf{y} = \sum_{i=1}^{d} g_i^2 \mathbf{p}_i^T \mathbf{p}_i + \mathbf{e}^{\mathrm{T}} \mathbf{e} \tag{13.20}$$

To terminate a learning process, we can measure the error reduction rate defined as

$$err = 1 - \frac{\sum_{i=1}^{k} g_i^2 \mathbf{p}_i^T \mathbf{p}_i}{\mathbf{y}^{\mathrm{T}} \mathbf{y}}, \forall k \in [2, d] \tag{13.21}$$

If $err \leq \varepsilon$, where $\varepsilon > 0$ is a small number, a learning process can be terminated with k selected independent variables. Equation (13.21) is similar to the definition of a normalised error defined in Chapter 7 if $\mathbf{y}$ is normalised with a zero mean and one standard deviation.

OLS can only be applied to regression problems. Here we use OLS to detect the relationships among ten top genes selected by the random forest algorithm in chapter 9. These top ten genes are BPSL2697, BPSS1512, BPSS0477, BPSL2522, BPSL2520, BPSL2096, BPSS1492, BPSL0326, BPSS2141, and BPSL3398. Ten OLS models are built. In each model, one of the genes is selected as the dependent variable while the rest are used as the independent variables. In each OLS model, we can analyse the weight vector **w** to see if any gene as independent variable dominantly contributes to other gene. It is found that only the genes BPSL2697 and BPSS0477 dominantly contribute to each other Fig. 13.8 shows the weight distributions of the two OLS models built using BPSL2697 (left panel) and BPSS0477 (right panel) as the dependent variable, respectively.

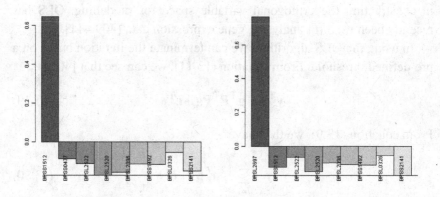

Fig. 13.8. Two OLS models built using BPSL2697 (left panel) and BPSS0477 (right panel) as the dependent variable. The horizontal axes represent the remaining nine genes as the independent variables in two models and the vertical axes represent the weights magnitudes.

13.4 Criteria for feature selection

There are two types of criteria in a feature selection process. One is to measure how good a sparse model is. When we add a new feature to a model, we need to measure how good the model is. For a regression application, the correlation between predictions and targets is one of the commonly used criteria. The errors between predictions and targets are also commonly used. In discussing the OLS algorithm, equation (13.21) is similar to the normalised error. In classification, classification accuracy or AUR can be used.

The second type is to determine which feature should be added into a model. Using an independent validation data set is an approach. For instance, when we have added a new feature to a model, we can re-estimate model parameters based on increased feature set. This newly estimated model is tested on the validation data set to see if the model performance is improved. Instead of using this empirical approach which introduces extra computational cost, the other approach is to measure how good a new feature is without using the validation data set.

To measure how good a feature is two factors need to be considered. First, can this new feature improve prediction power? Second, does this new feature bring unique contribution to the model compared with the selected features? To address these two questions quantitatively we introduce three metrics.

13.4.1 *Correlation measure*

In a regression application, correlation can be well used for measuring how one variable correlates with the other. When we add a new feature denoted by X, we can measure its correlation with the target variable denoted by Y, $\rho(X,Y)$. If we have a candidate set denoted by Θ, we need to maximise the correlation through

$$X_g = \arg\max_{X_i \in \Theta} \{ \rho(X_i, Y) \} \tag{13.22}$$

However, this new feature may not bring a unique contribution if it is highly correlated with the selected ones. This requires us to consider the second correlation measure. If the set of selected features is denoted by Ω, we need to consider

$$\arg\min_{X_i \in \Theta, X_j \in \Omega} \{ \rho(X_i, X_j) \} \tag{13.23}$$

In order to consider both measures for selecting a good new feature, we need to introduce a trade-off parameter $\alpha \in [0,1]$. Using this parameter we have

$$X_g = \arg\max_{X_i \in \Theta} \{ \alpha \times \rho(X_i, Y) \} + (1-\alpha) \times (1 - E[\rho(X_i, \Omega)]) \} \quad (13.24)$$

where $E[\rho(X_i, \Omega)]$ is the expected correlation of X_i with all selected features in Ω. It can be seen that the correlation between a newly selected feature and the dependent variable must be maximised while the correlation between the newly selected feature and the other selected features is penalised. If $\alpha = 0$, we select completely non-correlated features. If $\alpha = 1$, we select features no matter if they are correlated.

13.4.2 *Fisher ratio measure*

When conducting a classification project, correlation between an independent variable and a dependent variable which is discrete or binary may not be appropriate. In this case, the Fisher ratio which measures how separately two classes are using a feature can be used. If the Fisher ratio measure between X and Y is denoted by $\mathcal{F}(X,Y)$, equation (13.24) can be re-written as below for a classification project

$$X_g = \arg\max_{X_i \in \Theta} \{ \alpha \times \mathcal{F}(X_i, Y) \} + (1-\alpha) \times (1 - E[\rho(X_i, \Omega)]) \} \quad (13.25)$$

Note that the relationship between a newly selected feature and the selected feature is still measured using the correlation measure as both are normally numeric variables.

13.4.3 *Mutual information approach*

Correlation analysis is a linear approach. It is unable to measure nonlinear correlation between two variables. Here we introduce the mutual information approach which can be used to measure nonlinear correlation between two variables.

 Mutual information is the difference between the initial uncertainty and the conditional uncertainty. $X_k \in \Theta$ is a variable and $P(X_k)$ is the *a prior* probability. The initial uncertainty of X_k is measured when X_k is isolated (not selected) and is defined as

$$H(X_k) = -P(X_k)\ln P(X_k) \qquad (13.26)$$

Let $P(X_k \mid \mathcal{K})$ be the conditional probability of X_k given a class domain $\mathcal{K} = \{g_1, g_2, \cdots, g_C\}$. The conditional uncertainty measures the information of X_k given the class domain and is defined as

$$H(X_k \mid \mathcal{K}) = -\sum_{g_c \in \mathcal{K}} P(g_c)P(X_k \mid g_c)\ln P(X_k \mid g_c) \qquad (13.27)$$

The mutual information of X_k with the given class domain is then

$$\begin{aligned} I(X_k, \mathcal{K}) &= H(X_k) - H(X_k \mid \mathcal{K}) \\ &= \sum_{g_c \in \mathcal{K}} P(X_k, g_c)\log\frac{P(X_k, g_c)}{P(X_k)P(g_c)} \end{aligned} \qquad (13.28)$$

A new variable whose $I(X_k, \mathcal{K})$ value should be maximised for selection of features.

$$X_\circ = \arg\max_{X_k \in \Theta}\{I(X_k, \mathcal{K})\} \qquad (13.29)$$

Replacing $\mathcal{K}$ with Ω we have the other mutual information measurement for detecting the independence of a sequence under selection,

$$\begin{aligned} I(X_k, \Omega) &= H(X_k) - H(X_k \mid \Omega) \\ &= \sum_{X_l \in \Omega} P(X_k, X_l)\log\frac{P(X_k, X_l)}{P(X_k)P(X_l)} \end{aligned} \qquad (13.30)$$

where $P(X_k, X_l)$ is the joint probability between the selected variable, $X_l \in \Omega$, and a new variable for selection, $X_k \in \Theta$. A newly selected variable should satisfy

$$X_\circ = \arg \min_{X_k \in \Theta} \{I(X_k, \Omega)\} \tag{13.31}$$

In order to trade off between two measurements we have a selection criterion as below

$$J_M(X_k) = \alpha I(X_k, \mathcal{K}) - (1 - \alpha) I(X_k, \Omega) \tag{13.32}$$

where α is a constant. The constant is set at 0.7 favouring discriminant ability. We refer to J_M as the information gain. A newly selected variable therefore satisfies

$$X_\circ = \arg \max \{J_M\} \tag{13.33}$$

As a powerful feature selection approach, the mutual information approach has been widely used in bioinformatics projects [412-414].

Summary

This chapter has discussed three strategies for feature selection, i.e. the built-in strategy, the exhaustive strategy, and the heuristic strategy. Using the built-in strategy, the CART and the random forest algorithm are able to extract features for nonlinear models. Other algorithms can only be applied to model linear data. However, compared with CART and the random forest algorithms, these linear algorithms provide a simple interpretation for prediction purposes. The exhaustive strategy has the limitation of time complexity although it can be applied to data sets with small dimensionality. The heuristic strategy is the most widely used in the literature and has also been widely used in bioinformatics. In using the heuristic strategy, we need to carefully select an appropriate algorithm and a feature selection criterion.

Chapter 14

Feature Extraction
(Biological Data Coding)

To present a biological data set to a machine learning model, we are required to make sure that the data are representative, quantitative, and informative. This requires four focal points. First, the process must be as consistent as possible, i.e. providing invariant format and required resolution at any time. Second, the process must be as accurate as possible, i.e. if a new process is able to explore more information from biological data, it should replace the old one. Third, the process must be as effective as possible, i.e. taking into account the machine learning time cost. Fourth, the process should use as much biological knowledge as possible for the presentation of biological data. This process is similar to most applications in other disciplines and is called feature extraction. Note that feature extraction is often confused with feature selection. Feature selection is mainly to reduce noise in data by removing irrelevant variables in learning. However, feature extraction is to find a better way to present data to machine learning algorithms. Feature extraction is commonly a process done prior to feature selection. Unlike feature selection which is closely related to machine learning, feature extraction is closely related to subjects. Different disciplines will need different feature extraction approaches. This is why feature extraction is hardly a hot subject in machine learning. Rather, it is an important subject in some areas, such as image analysis, ECG signal processing, and sensor data analysis. We use a separate chapter for biological feature extraction to emphasise the importance of this topic in bioinformatics. In this chapter, the targets are molecular sequences and chemical compounds which are generally a chain of non-numerical components.

14.1 Molecular sequences

A DNA (Deoxyribonucleic acid) sequence is a chain of four nucleic acids, i.e. adenine, guanine, cytosine, and thymine. They are expressed by four letters, A, G, C, and T. A protein sequence is a chain of 20 amino acids shown in Table 14.1 where the full names, short names, and abbreviations are listed.

Table 14.1. Twenty amino acids.

Full name	Short name	Abb	Full name	Short name	Abb
Alanine	ala	A	Leucine	leu	L
Arginine	arg	R	Lysine	lys	K
Asparagine	asn	N	Methionine	met	M
Aspartic acid	asp	D	Phenylalanine	phe	F
Cysteine	cys	C	Proline	pro	P
Glutamine	gln	Q	Serine	ser	S
Glutamic acid	glu	E	Threonine	thr	T
Glycine	gly	G	Tryptophan	trp	W
Histidine	his	H	Tyrosine	tyr	Y
Isoleucine	ile	I	Valine	val	V

In proteomics, it is known that sequences determine structures and structures determine functions. Based on this, many structure and function prediction projects are studying sequence structures or sequence specificities for exploring the hidden relation between sequence specificities and protein structures and functions. In studying DNA sequences, it has also been found that the sequence specificity is closely related to genomic functions and organism speciation.

There are two different types of tasks in using sequence components for predictions. One is using whole sequences while the other is using short segments or peptides which are extracted from whole sequences. Studying whole sequences is a major focus in comparative genomics and comparative proteomics where the aim is to predict the structure and function of a whole molecule. For instance, we may need to investigate how splicing sites, translation start sites, methylation sites, and promotion regions are distributed in a new DNA sequence. We may also need to study how posttranslational modification sites, enzyme cleavage sites, or (or and) metal binding sites are distributed in a new protein sequence. To have a precise study, a wet laboratory experiment can be done. However, without *a prior* knowledge, a blind laboratory experiment can be very expensive and time consuming. In order to narrow down to the focal points, comparing a new sequence against some database sequences that have annotated structure and function information is a common approach used in bioinformatics. Various sequence homology alignment algorithms and tools are developed and implemented for this purpose. Discussing these algorithms and tools are beyond the scope of this book and the readers are recommended to read relevant textbooks [1-3, 5, 415]. However, we understand that sequence homology alignment algorithms and tools are mainly based on a database of annotated sequences. When this database is large, the computational cost is huge. For this reason, parametric models can be considered and feature extraction is needed.

Studying peptides is for investigating a single molecular function. For instance, we may need to study if a DNA segment has a methylation site, a splicing site, a translation start site, or a promoter region. For a protein, we may need to study whether a protein segment has a phosphorylation site, a hydroxylation site, a nitration site, or an enzyme cleavage site. In this case, most sequence homology alignment algorithms and tools are not appropriate. A proper feature extraction is then needed.

14.2 Chemical compounds

A chemical compound is defined as a chemical substance composed of two or more chemical elements such as oxygen, hydrogen, etc [416-418]. Each chemical compound has a unique structure, but can be decomposed through a chemical reaction. Chemical compounds are basic units in metabolism and most cellular functions. In viral and pathogenic studies, chemical compounds are important targets for drug design and testing.

A chemical compound can be expressed by a chemical formula which shows what chemical elements the compound is composed of and how many units of each chemical element are present. For instance, H_2O is a water compound, where two units of hydrogen and one unit of oxygen are used. NaCl represents a salt compound, containing one unit of sodium (Na) and one unit of chlorine. The expression of the chemical elements and units is called the Hill notation [419]. With the Hill notation, chemical elements have their order listed from left to right in a formula. In the periodic table, there are currently 117 chemical elements [420].

In order to study the relationships between chemical compounds and cellular functions, it is necessary to consider a proper approach to encode the chemical elements in chemical compounds.

14.3 General definition

If a whole sequence or a peptide is denoted by $s \in \Theta$ (Θ is a set of discrete states of values), a feature extraction process is expressed by

$$\mathcal{P} : s \mapsto x \in \mathfrak{R}^d \tag{14.1}$$

where $\mathcal{P}$ means a process and x is a coded vector. For instance, in handling DNA sequences or peptides, $\Theta = \{A, G, C, T\}$. The set of a protein sequence or peptide chain is expressed as $\Theta = \{A, C, D, E, F, G, H, I, K, L, M, N, P, Q, R, S, T, V, W, Y\}$. For a chemical compound, Θ is then a set of 117 chemical elements such as Oxygen, Hydrogen, Carbon, etc.

14.4 Sequence analysis

14.4.1 *Peptide feature extraction*

A peptide is commonly an extracted segment from a whole sequence. The length of peptides in a data set for analysis is commonly fixed. The study of peptides commonly focuses on the classification of peptides into different categories. Therefore, peptide data analysis is also called peptide classification. Meanwhile, the study of peptide data is for determining whether a certain function is involved. Peptide classification is therefore also known as functional site prediction. Given a set of N peptides as well as the labels of the peptides $\Omega = \{ \mathbf{s}_n, t_n \}_{n=1}^{N}$, the task is to build a classifier to map $\mathbf{s}_n$ to t_n

$$\mathcal{F} : f(\mathbf{s}) \mapsto t \tag{14.2}$$

where $\mathcal{F}$ is a family of functions while f is a specific function which can map $\mathbf{s}_n$ to t_n accurately. Using a machine-learning approach, we can try to find f in $\mathcal{F}$. In fact, a few machine-learning algorithms can handle non-numeric input data such as peptides. There is a need to convert peptides to numeric vectors before using machine learning algorithms as defined in equation (14.1).

The easiest way to code peptides or extract features from peptides is to use a sparse orthogonal coding approach [209] which has been widely used in bioinformatics. The approach uses a binary vector to represent each molecular basis, i.e. nucleic acid or amino acid. For four nucleic acids, the basic codes are 0001, 0010, 0100, and 1000. For 20 amino acids, the basic codes are 0000000000 000000001 for Alanine, 000000000 0000000010 for Cystein, etc. Based on these basic codes, a set of numeric features of a peptide can be extracted. For instance, the feature vector of a 4-mer protein peptide ACGT is 80 bits long. Within it there are four non-zero bits only.

This means that such a feature extraction approach generates very sparse feature vectors for peptide data. The advantages of this feature extraction approach are the simplicity and high resolution. It is easy to understand how simple this approach is. In terms of high resolution, we can imagine how data are sparsely distributed in a very high-dimensional space, where each data point is sitting in a corner of a hyper-cube if all the possibilities have been exhausted. Figure 14.1 demonstrates this distribution. If data points on the corners belong to different categories, a hyper-plane can be found to separate peptides as shown in Fig. 14.1 as a vertical plane.

However, the data space has been expanded so that it is unnecessarily large. A 4-mer protein peptide needs 80 independent variables and an 8-mer protein peptide needs 160 independent variables. A serious problem of this kind of data is that a feature selection may lead to a

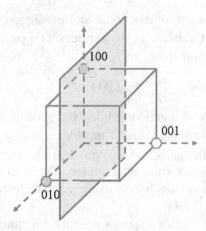

Fig. 14.1. A demonstration of sparse orthogonal coding as a feature extraction approach for peptide classification. Filled circles belong to one class while the open circle belongs to the other class. The vertical plane is an illustration of decision hyper-plane.

model hard to interpret because only a collection of every 20 consecutive bits (independent variables) makes sense for the interpretation. If part of these 20 bits is left after feature selection, it is not useful for interpretation.

The second approach is to use frequency estimation of molecular bases. The work of using frequency estimate in a computer program is done by the *h* function [421]. With the *h* function, the frequency of 20 amino acids at each residue is calculated from a set of functional training peptides. Each functional training peptide contains a functional site. The frequency estimate as a matrix with 20 rows for 20 amino acids and *k* columns for *k* residues in a peptide is then stored in a computer program. Such a matrix is referred to as a recognition rule. If the amino acids in a query peptide can hit a high frequency, the peptide will be considered as functional, otherwise non-functional. This approach is very simple and straightforward. However, the major shortcoming of this method is that is has a high sensitivity and a low specificity.

The third approach is to use various hydrophobicity scales for feature extraction for protein peptides. Seven hydrophobicity scales available in the literature have been used. They are the Kyte-Doolittle scale [163], the Hopp-Woods scale [422], the Cornette scale [423], the Eisenberg scale

[424, 425], the Engelman scale [426], the Janin scale [427], and the Rose scale [428]. The use of a hydrophobicity scale is due to its traditional role in analysing the impact of amino acid hydrophobicity on protein structure and potential for interaction and binding with other molecules [429]. Hydrophobic amino acids are generally located in the protein interior whereas hydrophilic amino acids are generally located on the protein surface as targets for binding with other molecules. A protein whose surface is composed of mainly negatively charged amino acids such as glutamate and aspartate will bind to a protein with mainly positively-charged residues such as lysine and arginine [430-434]. This means that the hydrophobicity scale is a candidate for encoding amino acids for constructing a predictive model. In using the hydrophobicity scales, there are two techniques for feature extraction. The first is to extract features for each peptide using a single value. The second is to use dual scales for feature extraction. This is because different hydrophobicity scales are developed based on different data in different laboratories. Difference is therefore seen among the seven scales. Some difference is large, i.e. a hydrophobic value in one scale can be a hydrophilic value in another scale.

An even more complicated feature extraction process can embed correlation between residues. Denote by $\phi_\omega(R_i)$ the code of the residue R_i using one hydrophobicity scale ω. We define $\phi_\omega(R_i)\phi_\omega(R_{i+g})$ as the correlation between two residues R_i and R_{i+g} with a gap $0 < g \leq \lambda$, where $\lambda < m$ is the maximum gap length which is pre-defined. For a peptide with length $m > g$, there will be $m - g + 1$ residue correlation measures. A mean value $E[\phi_\omega(R_i)\phi_\omega(R_{i+g})]$ can be taken. If $\lambda = 4$, there are four extra codes for a single-scale hydrophobicity pattern, namely $E[\phi_\omega(R_i)\phi_\omega(R_{i+1})]$, $E[\phi_\omega(R_i)\phi_\omega(R_{i+2})]$, $E[\phi_\omega(R_i)\phi_\omega(R_{i+3})]$, and $E[\phi_\omega(R_i)\phi_\omega(R_{i+4})]$. For a dual-scale hydrophobicity pattern, there will be eight extra codes expressed as $E[\phi_\omega(R_i)\phi_\omega(R_{i+1})]$, $E[\phi_\omega(R_i)\phi_\omega(R_{i+2})]$, $E[\phi_\omega(R_i)\phi_\omega(R_{i+3})]$, $E[\phi_\omega(R_i)\phi_\omega(R_{i+4})]$, $E[\phi_\tau(R_i)\phi_\tau(R_{i+1})]$, $E[\phi_\tau(R_i)\phi_\tau(R_{i+2})]$, $E[\phi_\tau(R_i)\phi_\tau(R_{i+3})]$, and $E[\phi_\tau(R_i)\phi_\tau(R_{i+4})]$, where $\tau \neq \omega$ is another hydrophobicity scale. Figure 14.2 shows the mechanism in this correlation feature extraction. When λ is increased, more correlation features will be introduced.

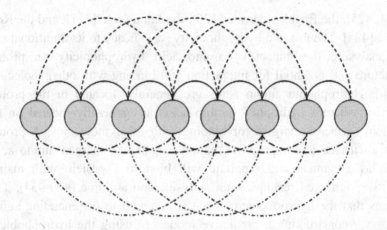

Fig. 14.2. An illustration of correlation feature extraction. The circles are the residues. The solid curves represent the correlation between two neighbouring residues without a gap. The broken curves represent the correlation between two neighbouring residues with a one-residue gap. The dashed curves represent the correlation between two neighbouring residues with a two-residue gap. The long dashed curves represent the correlation between two neighbouring residues with a three-residue gap.

The fourth method is based on a special learning mechanism we have studied in chapter 11, a basis neural network or a kernel machine. In such a machine learning algorithm, the original independent variables are not used as the direct input variables for a machine learning model. Instead, a kernel function is used to measure the similarity between data points. A model weights the contributions of various kernels, hence weighting the contributions of each training data point. These important training data points are called bases in basis function neural networks. In support vector machine, they are called support vectors while in relevance vector machine, they are called relevance vectors (in fact prototype vectors). Because the independent variables are not used as direct input variables to these models, feature extraction becomes an implicit process, i.e. a feature space is not the space of the original independent variables. A feature space is a space of kernels in this case.

Based on different kernel functions, the original data (peptide) space is mapped to a kernel space in which a machine learning algorithm works out how they contribute to a regression or a classification model. Here we discuss two interesting kernel functions.

Looking at the sparse orthogonal coding approach, we can recall the Needleman-Wunsch homology alignment algorithm [11], where we use a binary scoring system. In using a kernel function for the sparse orthogonal coding approach we end up with this simple homology alignment scoring system. Two peptides are denoted by s_i and s_j. If two peptides are coded by two binary vectors ($\mathbf{x}_i \in \{0,1\}^{20\times|s_i|}$ and $\mathbf{x}_j \in \{0,1\}^{20\times|s_j|}$) using the sparse orthogonal coding approach, one way to quantify their similarity is the dot product. The dot product between two vectors is defined as

$$\mathbf{x}_i \cdot \mathbf{x}_j = \sum_{k=1}^{|\mathbf{x}_i|} x_{ik} x_{jk} \qquad (14.3)$$

where x_{ik} and x_{jk} are the kth elements of $\mathbf{x}_i$ and $\mathbf{x}_j$ respectively. Because of the specificity of the sparse orthogonal coding approach for amino acids (or nucleic acids), equation (14.3) can be re-written as

$$\delta(s_{ir}, s_{jr}) = \begin{cases} 0 & if \ s_{ir} \neq s_{jr} \\ 1 & if \ s_{ir} = s_{jr} \end{cases} \qquad (14.4)$$

Here s_{ir} and s_{jr} are the rth residues of two peptides (s_i and s_j). This dot product actually is a special case of the Needleman-Wunsch scoring system. In support vector machine, it is called a linear kernel function

$$\sigma(s_i, s_j) = \mathbf{x}_i \cdot \mathbf{x}_j \qquad (14.5)$$

This kernel function can be further extended to a polynomial kernel function shown as below

$$\phi^P(s_i, s_j) = [\alpha\sigma(s_i, s_j) + \beta]^d \qquad (14.6)$$

where α, β, and d are the parameters of the polynomial function. For instance, the similarity between PRGLGPPG and LPGPGAPG is $\sigma(\text{PRGLGPPG, LPGPGAPG}) = 4$ and

$$\phi^P(\text{PRGLGPPG, LPGPGAPG}) = (\alpha 4 + \beta)^d \qquad (14.7)$$

In fact, this identity matrix is an extreme case of many mutation matrices. The Needleman-Wunsch algorithm, which was originally developed for molecular sequence homology alignment, has been replaced by many advanced algorithms like the Smith-Waterman algorithm [13] as well as some database sequence homology alignment tools like FASTA [14] and BLAST [10]. All of these new algorithms or tools use mutation matrices (the Dayhoff [15] score and its variants) rather than the identity matrix for scoring sequence similarity. The relationship between any pair of amino acids using the Dayhoff score is not *hard*. Instead, it becomes *softer*. The residue identity using a mutation matrix is then defined as

$$\sigma(s_{ir}, s_{jr}) = M(s_{ir}, s_{jr}) \qquad (14.8)$$

Here $M(s_{ir}, s_{jr})$ is a value from a mutation matrix. A relevant bio-basis function for using various mutation matrices to measure the similarity between two peptides has been introduced in chapter 11. Suppose two peptides are denoted by $\mathbf{s}_i \in \Theta^d$ and $\mathbf{s}_j \in \Theta^d$, where Θ is a set of nucleic acids or amino acids while d is the length of the peptides. The bio-basis function is defined as below [317, 318]

$$\phi^B(\mathbf{s}_i, \mathbf{s}_j) = \rho(-\beta\sigma(\mathbf{s}_i, \mathbf{s}_j)) \qquad (14.9)$$

where ρ is a sigmoid function.

14.4.2 *Whole sequence feature extraction*

When modelling whole sequence data, some feature extraction techniques mentioned in the last section are not applicable. The most widely used are the frequency features of k-mer motifs, where $k \geq 1$. A k-mer motif is a chain of k nucleic/amino acids. For instance, the features extracted using up to 2-mer motifs from a segment GCTCATTGCACTGCATTAAA can be shown in Table 14.2.

Table 14.2. The frequency features extracted from a DNA segment shown in the main text.

Motif	Frequency	Key	Frequency
A	6	CG	0
C	5	CT	2
G	3	GA	0
T	6	GC	2
AA	2	GG	0
AC	1	GT	0
AG	0	TA	1
AT	1	TC	1
CA	3	TG	2
CC	0	TT	2

The second feature approach for whole sequences is to count the frequency of certain types of nucleic/amino acids. For instance, amino acids can be classified in terms of physio-chemical properties [435-439] or Taylor classification [440], which is shown in Fig. 14.3. Eight features can be extracted using the Taylor classification of amino acids.

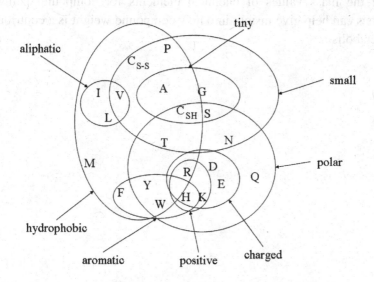

Fig. 14.3. The Taylor classification of 20 amino acids.

Summary

This chapter has discussed approaches for extracting features from molecular entities including DNA sequences, protein sequences, and chemical compounds. Feature extraction is an extremely important step in bioinformatics in three aspects. First, a proper feature extraction approach can explore as much hidden biological information as possible compared with an inappropriate approach which will not. The principle in machine learning is "garbage in, garbage out". If we present inaccurately extracted features for a machine learning algorithm, it will refuse to do us a favour. Second, an efficient feature extraction approach can save much computational time. There is a very general rule in machine learning, "the simplest is the beauty". This means that a simple machine learning model can generalise well compared to a complicated model. This is because a complicated model learns too much details (or noise) from data. Such an over-complicated model is often an over-fitted one. Third, a biologically-sound feature extraction approach can provide a good platform for interpreting a model. For instance, a model built using the mass values of chemical elements for compound pathway analysis can help give insight into how compound weight is a contributor in metabolism.

Chapter 15

Sequence/Structural Bioinformatics Foundation – Peptide Classification

In this chapter we discuss the foundation of sequence/structural bioinformatics through peptide classification. The discussion is conducted by covering two different applications. They are posttranslational modification site prediction and promoter region identification. Although they are very different in nature, the basic concept is to look at local regions (or segments or peptides) to study functionality of a molecule. The aim of the chapter is to demonstrate how to conduct independent bioinformatics research using machine learning algorithms through feature extraction, feature selection, model construction, model evaluation, and model selection.

15.1 Nitration site prediction

Tyrosine nitration is a newly discovered posttranslational modification (PTM) [441-446]. New studies have found that tyrosine nitration significantly affects signalling pathways for cellular signal transduction [447-451]. For instance, tyrosine nitration plays a key role in altering signal transduction during proinflammatory stress [452]. Other studies confirm that tyrosine nitration is the outcome of triggering signalling pathways by nitric oxide during NGF-induced neuronal differentiation in PC12 cells [453, 454]. As an important signal transduction activity, the mitogen-activated protein kinase (MAPK) signalling pathways can be manipulated by asbestos-induced tyrosine nitration [455]. Meanwhile the contribution of protein tyrosine nitration to signalling pathways triggered by nitric oxide has also been paid increasing attention [456-459]. For

225

instance, tyrosine nitration has been identified as a contributing biomarker of oxidative stress and the nitration of some tyrosine sites can modify protein functions. In medicine, identifying nitration pathways and nitrated proteins in disease states is highly related to and significantly contributes to human pathology studies [458]. Because of this, tyrosine nitration has been the target of a potential predictor of acute and chronic disease states [458]. In pharmaceutical research, tyrosine nitration has also been intensively studied. For instance, it has been found that tyrosine nitration is inhibited when using an anti-tubercular drug [460]. It is also shown that tyrosine nitration is linked with drug resistance in neuronal-like PC12 cells [453]. The drug called aminotetrahydrofuran derivative tetrahydro -N,N- dimethyl -5,5- diphenyl -3- furanmethanamine hydrochloride (ANAVEX1-41) has been found as a neuroprotective agent in Alzheimer's disease. The drug has been found to be able to prevent tyrosine nitration [461]. An experiment has found the relation between tyrosine nitration and the resistance of Doxorubicin while the drug has been used in cancer treatment [462].

There are 77 protein sequences with tyrosine nitration sites in the Swiss-Prot and 89 protein sequences in the NCBI database. Two data sets contain partially overlapped sequences. The extraction of tyrosine peptides follows a common practice in posttranslational modification site prediction model construction, i.e. forming a tyrosine peptide using symmetrically consecutive residues which flank every tyrosine in a protein sequence within a given window size. The evaluated peptide lengths (window sizes) are 10, 20, and 30. A tyrosine peptide is denoted by $N_m - X - N_1 - C_1 - X - C_m$ and C_1, where X means any residues, and N and C are used to denote the N-terminal and C-terminal residues, respectively. $R=2*m$ is used to denote the number of flanking residues (peptide length). A tyrosine peptide with an experimentally verified tyrosine nitration site in the middle (between N_1 and C_1) is labelled as positive (functional) while a tyrosine peptide which has not yet been confirmed as having a tyrosine nitration site in experiments is labelled as negative (non-functional). All the inferred tyrosine nitration sites are not used.

A duplication check is carried out separately for both data sets. Whenever a duplicated pair is found for two peptides belonging to the same category, i.e. being negative or positive, one is removed. It must be noted that there is a large possibility that a negative peptide is identical to a positive peptide. A tyrosine site which is not labelled as an experimentally verified tyrosine nitration site could be potentially a true tyrosine nitration site. If two tyrosine peptides of a duplicated pair belong to two different categories, the positive peptide is kept while the negative one is discarded. After this, two sets of tyrosine peptides are combined. During the combination process, one more duplication check process is conducted to ensure there are no identical peptides in the data. In order to have roughly balanced peptides from two categories for modelling, a random selection is used for negative peptides at this stage. Table 15.1 shows all the peptide information. For instance, in the 10-mer data set, 686 non-duplicated negative peptides and 42 non-duplicated positive peptides are extracted from 77 Swiss-Prot protein sequences. Meanwhile 718 non-duplicated negative peptides and 55 non-duplicated positive peptides are extracted from 89 NCBI protein sequences. After the combination, 73 negative and 56 positive 10-mer tyrosine peptides are maintained for hydrophobicity encoding. Two coding strategies are used. They are single-scale hydrophobicity patterns and dual-scale hydrophobicity patterns. A single-scale hydrophobicity pattern is generated using a single hydrophobicity scale for a tyrosine peptide while a dual-scale hydrophobicity pattern is generated using two different hydrophobicity scales, for instance the Kyte-Doolittle scale plus the Hopp-Woods scale. In total, there are 28 sets of hydrophobicity

Table 15.1. Peptide distribution for three peptide lengths and two peptide formation stages.

Peptide length	Swiss-Prot Protein	Swiss-Prot peptide	NCBI protein	NCBI peptide	Final peptide
10	77	686/42	89	718/55	73/56
20	77	699/42	89	722/56	71/57
30	77	703/42	89	725/56	68/57

patterns including seven single-scale ones and 21 dual-scale ones for each peptide. Residue correlations described in chapter 14 are used.

Four machine learning algorithms are used. They are classification tree, artificial neural network (ANN) [463], the support vector machine [333] and the random forest algorithm [272]. All are available in R. In using ANN, four model structures with four different numbers of hidden neurons (5, 10, 15, and 20) are constructed. The radial basis function is used for the kernel function in the SVM models. The cost function is 100 and the smoothing parameter of the radial-basis kernel function is one. The default parameters of the random forest model are used. The five-fold cross-validation approach [252] is adopted.

All classification tree models demonstrate much lower prediction accuracy than others (around 60%). ANN fails to model the following situations, 20-mer data with 20 hidden neurons, 30-mer data with 15 hidden neurons, and the 30-mer data with 20 hidden neurons because the ratio of the number of parameters over the number of data points in these models exceeded the limit set by the package. For each of three algorithms one top model is selected. Because there are different numbers of hidden neurons, there are more top ANN models compared with the other two algorithms. Through computer simulation, it is found that there is positive correlation between the specificity of tyrosine peptides and nitration status. Such positive correlation is then implemented as a predictive tool for tyrosine nitration site prediction.

Figure 15.1 shows the prediction performance of the top models ordered by the total accuracy. The ANN model with 10 hidden neurons built on the 30-mer data encoded by the Kyte and the Hopp scales is the best one. Its total prediction accuracy is 74%, sensitivity is 74%, specificity is 75%, negative prediction power is 73%, positive prediction power is 75%, and AUR is 0.75. The next best model is the RF model built on the 20-mer data encoded by the Kyte and the Hopp scales as well. Its total prediction accuracy is 73%, specificity is 75%, sensitivity is 72%, negative prediction power is 72%, positive prediction power is 75%, and AUR is 0.72.

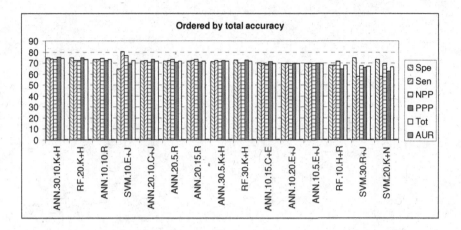

Fig. 15.1. The prediction performance ordered by the total prediction accuracy for the top models. The labels in the horizontal axis represent models. "ANN", "RF", and "SVM" are three machine learning algorithms. The numbers besides them indicate the peptide length. In each of nine top ANN models, the second number after the first number (for instance, 5, 10, 15, and 20) means the number of hidden neurons. "K" means the Kyte scale, "H" means the Hopp scale, "C" means the Cornette scale, "J" means the Janin scale, "R" means the Rose scale, "E" means the Eisenberg scale, and "N" means the Engelman scale. Dual scales are represented by two letters with "+" in between, for instance, "K+H" means the use of the Kyte scale and the Hopp scale. The vertical axis represents the percentage.

Figure 15.2 shows the ranking result (box plot) using the mean decrease in Gini gain based on five-fold cross-validation random forest models. It can be seen from the Figure that the eighth N-terminal flanking residue of a tyrosine encoded by the Kyte scale is the most important one for the prediction. Meanwhile all eight residue correlation codes are within the top half region. Among them, the residue correlation encoded by the Hopp scale with a gap of 3 is ranked the third and the residue correlation encoded by the Kyte scale with a gap of 4 is ranked the fifth. Two of the top five variables are residue correlation codes. This does illustrate the importance of residue correlation in predicting tyrosine nitration sites. Because there is little experimentally verified mechanism of tyrosine nitration so far, this ranking can give some clue for biological investigation.

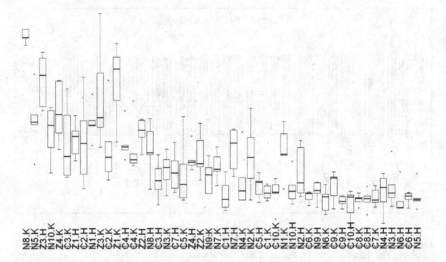

Fig. 15.2. Mean decrease in Gini gain for the RF model built on the 20-mer tyrosine peptides encoded by the Kyte scale and the Hopp scales. The variables are 40 residue codes (dual-scale codes for each of 20 residues) plus 8 residue correlation codes (4 for each of two scales). The data are ordered. The horizontal axis represents the encoded residues. "N" means an N-terminal residue and "C" means a C-terminal residue. The number following "N" or "C" means the residue number away from the tyrosine (the target for the prediction). The last letter represents the scale, K for Kyte scale and H for Hopp scale. "Z" means residue correlation codes. "Z4.K" means a residue correlation code with a gap length 4 (maximum gap length) and using the Kyte scale. A variable with a higher Gini gain is more important than a variable with a lower Gini gain. The vertical axis represents the Gini gain.

15.2 Plant promoter region prediction

A promoter is a segment with a few hundreds of nucleic acids of a DNA sequence. The function of a promoter is for facilitating gene transcription. Because of this, a promoter is located upstream of a DNA sequence near a gene. A promoter covers about 200 base pairs upstream and 51 base pairs downstream of a transcriptional start site. The close relation between promoters and gene transcription has made promoters an important contributor to various cellular functions including disease development. The prognostic importance of promoters has therefore already been discovered for various diseases [464-467].

In prokaryotic organisms, it is believed that there is a conserved motif in the region of -35 base pairs upstream of a transcriptional start site [367, 468-470]. Predictors have been developed for prokaryotic promoter region prediction [471-473]. There is a large diversity when characterising promoters in eukaryotic proteins. A few kilo-base pairs upstream of a transcriptional start site may need to be examined [474]. Although it is accepted that eukaryotic promoter region prediction is difficult [475-478], a number of predictors have been developed. For instance, self-organising map [145] is used for analysing nucleic acid profile in promoters [479]. Artificial neural network is used for promoter prediction [477]. AdaBoost is used for the prediction of promoters as well [478].

Plant promoters, which are the focus of this work, have a close relation to pathogens. For instance, it has been found that an avirulence protein (for instance, AvrBs3 or AvrBs3Deltarep16) binds and activates a promoter of disease resistance genes to fight against pathogen invasion [480]. *Xanthomonas campestris* pv. Campestris is known as a causal agent of black-rot disease of cruciferous plants. It is related to the *Escherichia coli* lac promoter [481]. In studying Rab/GTPases which can regulate vesicular trafficking during exocytosis, endocytosis and cellular differentiation, it is found that a pectinase gene promoter allows foreign genes on pectin medium [482]. WRKY factors, known as a family of plant-related transcriptional regulators, are related to plant stress control. In an experiment, it is found that WRKY can suppress its own promoter activity and is positively correlated with pathogen defense-associated PR1 promoter activity [483]. Some synthetic promoters have been produced to study how signalling and transcriptional activation function when plant-pathogens are injected [484].

In terms of the importance of plant promoters, a database called PlantProm has been recently established [485]. In the database, all plant promoters for polymerase II are annotated without redundancy. All promoters have experimentally verified transcriptional start sites. There have been 175 TATA-rich plant promoters and 130 TATA-less plant promoters annotated so far. All the promoter segments are composed of -200 upstream base pairs of the transcriptional start sites and +51 downstream base pairs of the transcriptional start sites. All segments are stored in the FASTA format. Based on this database, two predictors have recently been constructed. One employed the transductive confidence machine [486] and the other employed the support vector machine algorithm [487]. In the former work, the sequence content and signal features are used as features. In the latter paper, the 4-mer motifs are used as features. Both have achieved very good prediction performance. However, motif correlation in relation to plant promoter has not drawn much attention.

175 TATA-rich plant promoters (sequence segments) and 130 TATA-less plant promoters are downloaded from PlantProm [485]. These data are treated as positive data. Each segment has 251 base pairs (bps) with -200 upstream and +51 downstream of transcriptional start sites. In order to generate negative data for model construction, 23211 nucleic acid sequences containing CDS are downloaded from NCBI [488]. This is a method used in the previous studies [486, 489]. For each of 23211 sequences, there is often more than one CDS segment. Among a number of CDS segments, one CDS is randomly selected. One segment of 251 bps is randomly selected from the selected CDS. Because not every sequence contains a CDS segment, only 20925 segments of 251 bps are extracted from these 23211 sequences. For each segment (both positive and negative), motif frequencies and motif correlations are used as features. The data composed of motif features are then used for building predictors.

Three sets of motifs are investigated. The first set targets sequence content, i.e. 1-mer, 2-mer, 3-mer and 4-mer motifs are used. This is to test how motif frequency is related to the recognition of a promoter region. The second set is called first-order or low-order motif correlation. It targets base pair correlation, for instance one adenine finds another adenine with a gap g. This is to evaluate how correlated single base pair motifs are related to promoter binding. The maximum value of g is set at 20. The third set is an extension of the second set aiming to investigate high-order motif correlations in relation to promoter status. Three machine learning algorithms: artificial neural networks, the support vector machine algorithm and the random forest algorithm, are used.

Three types of motifs are designed. Type I are k-mer motifs. With this motif pattern, a sliding window with a target k-mer motif is used to scan each segment. The frequency of a k-mer motif in a segment is used as a feature. When k is 1, this is single nucleic acid frequency, i.e. the frequencies of cytosine, guanine, adenine and thymine in a segment. Here maximum k is set to 4. In total, there are 340 type I motifs, hence 340 features.

Type II motifs are called first-order or low-order motif correlation, i.e. gapped single-nucleic correlations. This measures how likely an adenine is to meet another adenine with a gap g within the same segment or how likely an adenine is to meet a thymine with a gap g within the same segment. The frequency of each gapped single-nucleic correlation is treated as a feature. Suppose we denote two of all the gapped single-nucleic correlations as A-1-A and A-1-T; the frequency of A-1-A and A-1-T is 1 and 2 in a segment GCTCATTGAACTGAATAAA. The maximum value of gap g is set to 20. The number of extracted features using Type II (or low-order) motifs is then 320=16*20.

Type III is an extension to Type II, i.e. being high-order motif correlations. The correlations of 2-mer and 3-mer motifs are evaluated in this study. Higher order motifs made model construction infeasible because of huge dimensionality. For instance, in a segment GCTCATTGCACTGCATTAA, the GC pair has one hit by a gap 3 and one hit by a gap 6. The TT pair has one hit by a gap 8. The CA pair has two hits by a gap 3. The triple ATT has one hit by a gap 7. The maximum gap is set to 20 in this study. There are 1600 high-order motif correlation features, i.e. (16+64)*20. The separate consideration of motif correlation is for investigating if low-order or high-order motif correlations are important in plant promoter recognition.

Three simulations are designed, i.e. **alpha** simulation, **beta** simulation and **gamma** simulation. In *alpha simulation*, TATA-rich and TATA-less segments are treated separately, i.e. two separate sets of models are constructed. One uses the TATA-rich segments as the positive data and the other uses TATA-less segments as the positive data. Both use the segments extracted from CDS in NCBI sequences as the negative data. In *beta simulation*, both TATA-rich and TATA-less segments are combined together to form the positive data while the segments extracted from CDS in NCBI sequences are used as the negative data, as previous works have done [486, 489]. The segments being organised this way is referred to as **dual** segments. Both *alpha* and *beta* simulations employ the 20-fold cross-validation approach for model evaluation [252]. In 20-fold cross-validation, data are randomly divided into 20 folds. One fold of data is reserved for testing using a model constructed on the rest of the data (19 folds). The 20 folds are in turn used as testing data. 20 models are therefore constructed. The mean and standard deviation of testing performance of the 20 models are calculated for the evaluation. In *gamma simulation*, three data sets are formed. Their positive data are TATA-rich segments, TATA-less segments and dual segments. Their negative data are the segments extracted from CDS in NCBI sequences as above. A bootstrapping approach is adopted to test the probabilistic blind sensitivity (**PBS**), i.e. estimating the probabilistic property when using constructed models to predict true promoters on novel segments. In order to test this, a certain proportion of positive segments (TATA-rich, TATA-less or dual segments) are randomly selected from whole positive segments as the **blind** segments. This simulation is designed for comparing with Shahmuradov's work using the transductive confidence machine [485] where only one set of blind segments is used for the evaluation. The rest of the data including positive and negative segments are treated as training data used for constructing a predictor. The numbers of reserved blind segments are 40 for TATA-rich (as in Shahmuradov's work [485]), 25 for TATA-less (as in Shahmuradov's work [485]) and 40 for the dual data. The process is repeated 100 times, i.e., 100 sets of blind segments are randomly generated from whole positive data. A predictor is constructed for each randomly generated set of training data by the 20-fold cross-validation approach. Each built predictor is then tested on its corresponding blind segments. A probabilistic estimation (mean and standard deviation) of blind sensitivity is obtained from the 100 random models.

In Shahmuradov's paper [486], two models are constructed for predicting TATA-rich promoters and TATA-less promoters in plants, separately. In prediction, 40 TATA-rich promoter segments and 25 TATA-less promoter segments are reserved as test segments and the rest are used for model construction. No negative data are used for testing the predictor. The prediction accuracies are 87.5% and 84% for TATA-rich and TATA-less, respectively. In Anwar's study [489], both TATA-rich and TATA-less promoters are combined into one model. The prediction specificity is 90% and the prediction sensitivity is 86%.

In comparison, the ANN models are the best. In all simulations, motif correlation (high-order motif) ANN models outperform others. This means that motif correlation plays an important role in promoter recognition. The performances of ANN models for the *beta* simulation are discussed here.

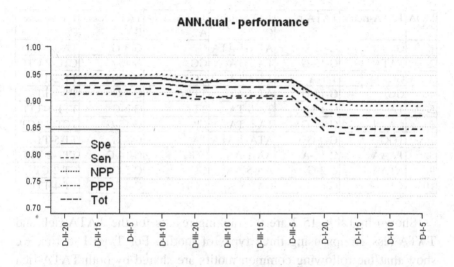

Fig. 15.3. The performance of ANN-dual model in alpha simulation. The notation of the horizontal axis follows TATA-coding type-hidden neurons. For instance, D-III-20 means dual data using Type III coding and an ANN model with 20 hidden neurons. The vertical axis represents the performance.

Figure 15.3 shows the performance of the alpha simulation for the dual group. The specificity of 12 models varies from 90% to 94%. The sensitivity of 12 models varies from 83% to 92%. All ANN models have their specificity greater than that of Anwar's work. 15 out of 18 ANN models have their sensitivity greater than that of Anwar's work. NPP varies from 89% to 95%. PPP varies from 85% to 91%. The total accuracy varies from 87% to 93%. The top model uses Type I coding with an ANN model employing 10 hidden neurons.

One of the important functions that the random forest algorithm has is to rank variables. Here we use the mean decrease in accuracy (MDA) for the analysis.

Table 15.2. The ranking results using MDA for two groups with three types of motifs. For Type II motifs, the features are expressed by X-n-Y, where X and Y are two nucleic acids and n is the number of gaps between them. For Type III motifs, the features are expressed by [X]-n-[X], where X is a motif and n is the gaps between two positions of the motif.

MDA	TATA-rich - A	TATA-rich - B	TATA-rich - C	TATA-less - A	TATA-less - B	TATA-less - C
1	G	G-12-G	ATA-4-ATA	G	G-3-G	TA-17-TA
2	TATA	G-15-G	TA-2-TA	GG	G-15-G	CTC-2-CTC
3	GGA	T-1-A	TA-4-TA	GGA	G-12-G	TT-2-TT
4	GG	G-18-G	AT-4-AT	TTTT	C-2-G	TT-1-TT
5	TA	G-1-G	TA-8-TA	TTT	G-6-G	TTT-1-TTT
6	GA	G-17-G	TA-6-TA	GA	G-18-G	TCT-2-TCT
7	TGG	G-9-G	ATA-2-ATA	TA	C-5-G	TT-5-TT
8	TAAA	C-15-A	AA-1-AA	TGG	G-14-C	TTT-2-TTT
9	ATAA	C-2-G	AA-5-AA	GC	C-5-C	CT-2-CT
10	TAA	G-6-G	CA-4-CA	CCTC	C-4-G	TT-4-TT

Shown in Table 15.2 are the ranking results for the TATA-rich and TATA-less groups using three types of motifs. For Type I motifs, we show that the following common motifs are shared by both TATA-rich and TATA-less groups, G, GG, TA, GA, and TGG. As expected, TATA-rich is highly ranked in the TATA-rich group, but not in the TATA-less group. Instead, the motif TTTT is highly ranked in the TATA-less group. The binuclear GC has been ranked at 9th in the TATA-less group, but not in the top ten of the TATA-rich group. When using Type II motifs, it is found that G-12-G, G-15-G, G-18-G, C-2-G, and G-6-G are common to the two groups. T-1-A, G-1-G, G-17-G, G-9-G, and C-15-A are distinct for the TATA-rich group while G-3-G, C-5-G, G-14-G, C-5-C, and C-4-G are distinct for the TATA-less group. When using Type III motifs, we show that the TATA-less group prefers rich thymine-based motif correlation. All top ten motif correlation features involve thymine. Moreover, four 2-mer motifs involve only thymine (TT-1-TT, TT-2-TT, TT-4-TT, and TT-5-TT) and two 3-mer motifs involve only thymine

(TTT-1-TTT and TTT-2-TTT). In the TATA-rich group, more motifs involve thymine and adenine.

Summary

This chapter has presented two case studies for understanding the foundation of sequence/structural bioinformatics. The core of these studies is peptide classification although they have different biological backgrounds. In peptide classification, we need to carefully collect sequence data and extract peptides. The duplicated ones must be removed. Care should also be taken for the inferred ones because they cannot be treated as annotated data or non-annotated data. These data must be excluded from the study in case they bring in some false information. In model construction, we need to be careful about model selection and evaluation. Different criteria may lead to the selection of different models. In most cases, a more robust model is preferred. Model selection should also be based on the evidence collected in a study. It is very common that one machine-learning algorithm outperforms the others in one application, but may show a completely different story in other applications. Making a web tool for public use is also a recommended research approach.

Chapter 16

Gene Network – Causal Network and Bayesian Networks

A traditional point of view of structure-function relation is

Sequence $\rightarrow$ Structure $\rightarrow$ Function

It is argued that this simple flow is too reductionistic [490]. This is because any biological function is completed by a network of molecules in a cell. A network is always composed of complicated interactions among molecules in a cell. Even more complicated are the dynamics of the interactions, some being more transient and some taking more time to complete. The complexity can increase when crosstalk happens unexpectedly. The post-genomic approach towards studying biological functions has a new flow expression as below [490]

Interaction $\rightarrow$ Network $\rightarrow$ Function

A gene network is a typical one among various molecular networks. This chapter focuses on the topic of causal network construction and the concept of Bayesian network.

16.1 Gene regulatory network

A gene regulatory network (GRN) is a graphical or visual representation of the interactions of DNA segments in a cell. Such a network has three properties. First, input and output; no GRN is an isolated entity in a cell. A GRN is activated by signals sent by other GRNs or external stimuli and sends signals to other GRNs. For instance, a plant-pathogen GRN is shown in Fig. 16.1, where the inputs to the network are rhizobacteria, fungus, bacteria including AvrPto, elicitor, and AvrRPM1. The output is the phytoalexins. A simple input-output function between them can be expressed as phytoalexins = f(rhizobacteria,AvrPto,elicitor,AvrRPM1)   (16.1)

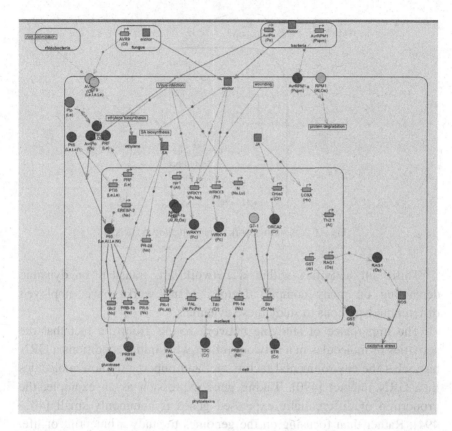

Fig. 16.1. Plant-pathogen gene regulatory network http://wwwmgs.bionet.nsc.ru/mgs/gnw/genenet/viewer/Plant-pathogen.

Second, the molecules in a GRN can have complicated interactions. Figure 16.2 is a sub-network of Fig. 16.1, where it can be seen that Pto (a kinase conferring resistance to tomato bacterial speck disease and interacting with proteins that bind a cis-element of pathogenesis-related genes [491]) receives both external signals (AvrPto) and signals from the nucleus (PRF). In return, it sends signals to the nucleus again through Pti5.

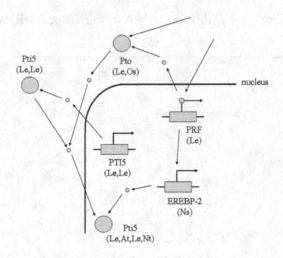

Fig. 16.2. A sub-network of plant-pathogen GRN.

Third, all activities within a network are transient or dynamic depending on many complex factors. Many works have employed differential equations to study these dynamics.

The importance of studying network results from the fact that the activities of molecules in a network reflect what initial conditions a GRN has, what the environmental factors are, and importantly how molecules in a GRN interact [490]. Taking gene expression as an example, the proportion of differentially expressed genes is commonly small [492-494]. Rather than focusing on the genomes to study a blueprint of life, studying how these differential genes appear is a way to discover the chemical blueprint of life through network analysis.

As mentioned above, a GRN is a graphical or visual representation of a true biological network. A GRN resembles most ordered networks for which there are two types of tasks for research. One is how to analyse network property and the other is how to construct a network based on observations. For the former, the focal points include network complexity analysis [495-497], shortest/longest path identification [498-500], and network robustness [501-503]. Machine-learning algorithms have been largely used for the latter. Among various algorithms, the Bayesian network shows many advantages in applications.

16.2 Causal networks, networks, graphs

A causal network is a network with connected nodes and arcs demonstrating causality. Each node represents a variable and each arc represents the causality between two variables. Variables with a direct connection are said to have a direct causality. Variables with an indirect connection are said to have an indirect causality. Variables having no connections are said to be independent. For instance, variables A and B have a direct causality in Fig. 16.3. Variables A and E have an indirect causality. Variables F and G are independent.

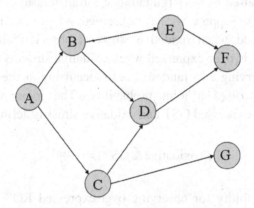

Fig. 16.3. An illustration of a causal network with six variables and causalities.

Because a causal network is a visual representation of a physical or biological system, a graph is commonly used to denote a network. A graph is a two-fold set $\mathcal{G} = (\mathcal{V}, \mathcal{E})$ with $\mathcal{V}$ to represent a set of variables (or nodes) and $\mathcal{E} \subseteq \mathcal{V} \times \mathcal{V}$ to represent a set of arcs. $(v_i, v_j) \in \mathcal{E}$ represents a directed arc from $v_i \in \mathcal{V}$ to $v_j \in \mathcal{V}$. In this book, only directed networks are used for the discussion. The notation $< v_{1'}, v_{2'}, \cdots, v_{n'} >$ is used to denote a path from $v_{1'}$ to $v_{n'}$ through variables $\{v_{2'}, \cdots, v_{n-1'}\}$. The notation $\hat{v}$ is used to represent the parent node(s) of a node v and the notation $\tilde{v}$ is used to represent the child node(s) of a node v.

16.3 A brief review of the probability

Informally, a probability is defined as a chance that an event happens among a discrete set of events. Suppose we have an experiment with N trials for testing oxidative stress in plants. The chance that the oxidative stress is observed in N experiments is expressed as

$$P(\text{oxidative}) = \frac{\text{the times that oxidation is observed}}{N} \qquad (16.2)$$

It is treated as a probability. In Fig. 16.1 we have seen that the oxidative stress can be caused by GST (glutathione S-transferase) or ROS (reactive oxygen species). Suppose we have observed $N_{\text{GST}} < N$ times that GST is over-expressed when oxidative stress is observed and $N_{\text{ROS}} < N$ times that ROS is over-expressed when oxidative stress is observed. The chance of observing two random events occurring at the same time in probability is called a joint probability. The joint probability for observing over-expressed GST and oxidative stress is defined as

$$P(\text{oxidative \& GST}) = \frac{N_{\text{GST}}}{N} \qquad (16.3)$$

The joint probability for observing over-expressed ROS and oxidative stress is defined as

$$P(\text{oxidative \& ROS}) = \frac{N_{\text{ROS}}}{N} \qquad (16.4)$$

Having these joint probabilities on hand, we may need to answer a question. What is the probability that GST will cause oxidative stress and what is the probability that ROS will cause oxidative stress? This is a question of conditional probability, i.e. the probability of one random event happening after another random event has happened. To answer this question, we need to know two more quantities: the probability that GST is over-expressed and the probability that ROS is over-expressed. Suppose they are also observed in this experiment and are denoted by P_{GST} and P_{ROS}. They are called marginal probabilities. Having these two more quantities, we can answer the above question using the concept of conditional probability. The conditional probability is defined as

$$\text{conditional probability} = \frac{\text{joint probability}}{\text{marginal probability}} \tag{16.5}$$

Going back to the question, we can see that the conditional probability that oxidative stress occurs with over-expression of GST is

$$P(\text{oxidative} \mid \text{GST}) = \frac{N_{\text{GST}}}{N \times P_{\text{GST}}} \tag{16.6}$$

and the conditional probability that oxidative stress occurs with over-expression of ROS is

$$P(\text{oxidative} \mid \text{ROS}) = \frac{N_{\text{ROS}}}{N \times P_{\text{ROS}}} \tag{16.7}$$

Figure 16.4 shows a diagram of the relationship between the three parts.

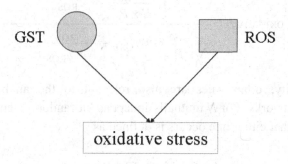

Fig. 16.4. The logical relationship between the oxidative stress, GST and ROS.

Remember that all the probabilities calculated above are based on the current experiment. Equations (16.6) and (16.7) are also called likelihood. They are objective measures. Before the experiment we may have some subjective knowledge about how likely GST and ROS are to be over-expressed, let's use another two probabilities to quantify this, i.e. π_{GST} and π_{ROS}. They are called *a prior* probabilities. Based on these probabilities, we can update our belief about which causes the oxidative stress using the Bayes rule. It is used here to deliver posterior probability for decision making

$$\text{posterior} = \frac{\text{conditional} \times \text{a prior}}{\text{evidence}} \tag{16.8}$$

The evidence is calculated as the sum of all the products between conditional probabilities and *a prior* probabilities. Going back to our question, we can see that the posterior probability that GST causes the oxidative stress is

$$P(\text{GST} \mid \text{oxidative}) = \frac{\dfrac{N_{\text{GST}}}{N \times P_{\text{GST}}} \times \pi_{\text{GST}}}{\dfrac{N_{\text{GST}}}{N \times P_{\text{GST}}} \times \pi_{\text{GST}} + \dfrac{N_{\text{ROS}}}{N \times P_{\text{ROS}}} \times \pi_{\text{ROS}}} \tag{16.9}$$

The posterior probability that ROS causes the oxidative stress is

$$P(ROS \mid \text{oxidative}) = \frac{\dfrac{N_{\text{ROS}}}{N \times P_{\text{ROS}}} \times \pi_{\text{ROS}}}{\dfrac{N_{\text{GST}}}{N \times P_{\text{GST}}} \times \pi_{\text{GST}} + \dfrac{N_{\text{ROS}}}{N \times P_{\text{ROS}}} \times \pi_{\text{ROS}}} \tag{16.10}$$

In probability, other rules are also relevant to the analysis of the Bayesian networks. For N mutually independent random events (x_i), the probability that either one occurs is defined as

$$P\left(\bigcup_{i=1}^{N} x_i\right) = \sum_{i=1}^{N} P(x_i) \tag{16.11}$$

The probability that all N mutually independent random events (x_i) occur at the same time is defined as

$$P\left(\bigcap_{i=1}^{N} x_i\right) = \prod_{i=1}^{N} P(x_i) \tag{16.12}$$

If two mutually independent random variables X and Y have N implementations (values), the total probability is defined as

$$P(x_i) = \sum_{j=1}^{N} P(x_i, y_j) \qquad (16.13)$$

or

$$P(y_i) = \sum_{j=1}^{N} P(y_i, x_j) \qquad (16.14)$$

16.4 Discrete Bayesian network

In this book, we provide only an introduction to discrete Bayesian network. For a variety of Bayesian networks, readers can refer to Neapolitan's book [504] and Jensen's book [505]. A Discrete Bayesian network (DBN) is denoted by $\mathcal{N} = (X, \mathcal{G}, \mathcal{P})$, where X represents a set of discrete variables, $\mathcal{G}$ represents a graph, and $\mathcal{P}$ represents a set of conditional probabilities. For each variable in $X_i \in X$, its probability distribution is denoted by $P(X_i \mid \hat{X}_i) \in \mathcal{P}$. Let's discuss the case shown in Fig. 16.4. In DBN, we assume that all three variables are taking binary values $\{$no, yes$\}$. The conditional probabilities $\mathcal{P}$ are shown in Table 16.1.

Table 16.1. The conditional probability set for the case shown in Fig. 16.4.

GST	ROS	Oxidative	
		no	yes
no	no	0.7	0.3
no	yes	0.2	0.8
yes	no	0.1	0.9
yes	yes	0.4	0.6

We also have the *a prior* probability set shown in Table 16.2. These probabilities quantitatively measure the subjective knowledge.

Table 16.2. The prior probability
set for the case shown in Fig. 16.4.

	π_{GST}	π_{ROS}
no	0.7	0.8
yes	0.3	0.2

The calculated posterior probabilities are

$$P(\text{GST} \mid \text{oxidative}) = (0.526, 0.474) \qquad (16.15)$$

and

$$P(R\textit{OS} \mid \text{oxidative}) = (0.722, 0.278) \qquad (16.16)$$

We can label the DBN by these numbers shown in Fig. 16.5, where the broken lines represent "no" state and the solid lines represent "yes" state.

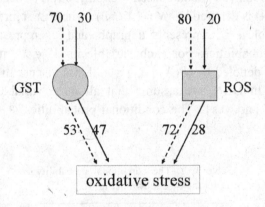

Fig. 16.5. The labelled DBN.

16.5 Inference with discrete Bayesian network

Building a machine-learning model is ultimately for prediction or inference. Given a Bayesian network, the aim is to predict the future with uncertainty. In a given DBN, we are required to calculate the marginal probability of a target variable using

$$P(Y) = \sum_{X \in X \setminus Y} P(X) = \sum_{X \in X \setminus Y} \prod_{X_i \in X} P(X_i \mid \hat{X}_i) \qquad (16.17)$$

In the oxidative stress case, if the target variable is oxidative, its parent variables are GST and ROS. We then have

$$P(\text{oxidative})$$
$$= \sum_{\text{GST}} \sum_{\text{ROS}} P(\text{oxidative}|\text{GST,ROS})P(\text{GST})P(\text{ROS}) \qquad (16.18)$$

Using the above equation, the marginal probabilities of oxidative stress are 47% and 53%. The probabilities indicate that it is very likely that oxidative stress is observable in this case.

16.6 Learning discrete Bayesian network

To estimate the parameters of DBN, the Bayes rule is used. Equation (16.8) can be re-written as

$$P(\vartheta \mid \mathcal{D}) = \frac{P(\mathcal{D} \mid \vartheta)P(\vartheta)}{P(\mathcal{D})}, \forall \vartheta \in \Theta \qquad (16.19)$$

where Θ is a parameter space and ϑ is the parameter learnt for fitting the data $\mathcal{D}$. In general, we assume that all model parameters are independent. Based on the probability rule discussed above, we have

$$P(\vartheta) = \prod_{X_i \in X} \prod_{\hat{X}_i \in X} P(\vartheta_{X_i|\hat{X}_i}) \qquad (16.20)$$

where $P(\vartheta_{X_i|\hat{X}_i})$ is called a local parameter prior for variable X_i which can use the Dirichlet distribution [506].

16.7 Bayesian networks for gene regulatory networks

Bayesian networks have been widely used in gene regulatory network analysis [507-513]. However, learning a Bayesian network model has been known to be an NP-hard problem, i.e. it is only applicable to small size networks. Searching and scoring approaches as well as a number of heuristic algorithms are implemented [514] [515] [516]. Meanwhile, conditional independency has to be well calculated. In a modified Bayesian network, mutual information is proposed to measure the dependency. Because of this, the exponential time complexity spent in calculating conditional independency can be reduced for large-scale gene regulatory network analysis [517]. In order to improve inference confidence, the idea of consensus and meta-analysis is used in analysing regulatory gene network [518]. Bayesian networks can also be used to reveal dynamics in gene expression data using time series gene expression data [519].

16.8 Bayesian networks for discovering peptide patterns

Here we demonstrate an application of discrete Bayesian networks for discovering patterns in peptide data. First, we apply DBN to the HIV data which was described in chapter 11. In applying DBN, the cleaved (positive) and non-cleaved (negative) peptides are separately used for building DBN models. After the modes have been constructed, we then visualise and compare two network structures. Figure 16.6 shows two network structures. The two structures have some differences. In the negative structure, it can be seen that eight residues are divided into four coupling groups regularly. The coupling effect is a common phenomenon in peptide structure analysis, i.e. neighbouring residues can possess certain structures for preserving biological pattern for certain biological functions. In the positive structure (the right panel), such a regular structure does not exist. Instead, residue C_4 and C_1 become parent variables for the other four residues.

Figure 16.7 shows a case of Hepatitis C-virus 10-mer protease peptides. The data is from an early study [520]. The data set is composed of 752 non-cleaved HCV peptides and 168 cleaved peptides. The same approach used in analysing the HIV data applies to the HCV data. The left panel is for the negative structure where the regular pattern is shown again. However, in the positive DBN structure (the right panel), such regularity is broken. The C_3 residue is not coupling with C_4. Instead, it correlates with N_1. Residue C_2 is the parent variable for C_1 and N_2.

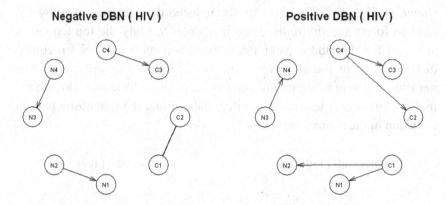

Fig. 16.6. Two DBN structures made for the HIV data. The left panel shows the negative structure and the right panel shows the positive structure. Each 8-mer peptide is denoted by N_4-N_3-N_2-N_1-C_1-C_2-C_3-C_4 with N meaning the N-terminal residues and C meaning the C-terminal residues.

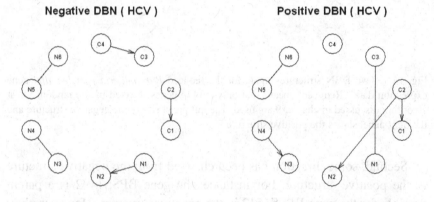

Fig. 16.7. Two DBN structures made for the HCV data. The left panel shows the negative structure and the right panel shows the positive structure. Each 10-mer peptide is denoted by N_6-N_5-N_4-N_3-N_2-N_1-C_1-C_2-C_3-C_4 with N representing N-terminal residues and C representing C-terminal residues.

16.9 Bayesian networks for analysing *Burkholderia pseudomallei* gene data

The positive and negative data sets (for infected and non-infected patients respectively) are used separately. The networks derived are shown in Fig. 16.8. The data are the reduced data set generated by the random forest algorithm discussed in chapter 9. Only the top ten genes are used for the study. After the construction of two DBN structures, differences were found between the two. First, the complexity in the negative structure is higher than that in the positive structure. This means that randomness is less in the positive data or that the diagnostic pattern is hidden in the positive structure.

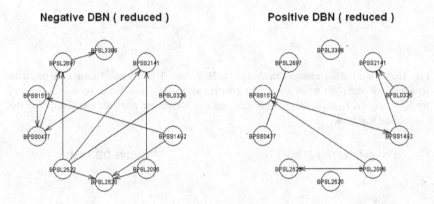

Fig. 16.8. Two DBN structures made for the reduced *Burkholderia pseudomallei* gene expression data. "Reduced" means that only ten top genes selected by the random forest algorithm discussed in chapter 9 are used. The left panel shows the negative structure and the right panel shows the positive structure.

Second, some direction has been changed from the negative structure to the positive structure. For instance, the gene BPSS1492 is a parent variable for the gene BPSS1512 in the negative structure. However, their relation in the positive structure is reversed. Third, one important mutual relation discovered in chapter 13 is retained for both negative and positive structures. It is the correlation between the gene BPSL2697 and the gene BPSS0477. Fourth, the gene BPSL2096 is always the root variable or the most causative gene in both negative and positive structures. The paths led by the gene BPSL2096 are shown in Fig. 16.9. In two paths, one "downstream" gene is identical, i.e. the gene BPSS2141.

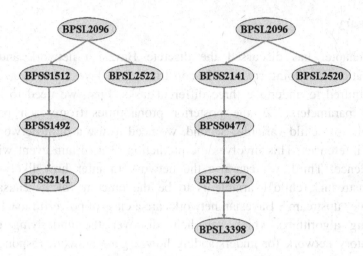

Fig. 16.9. Two paths led from the same gene (BPSL2096) from both negative (the right panel) and positive (the left panel) DBN structures.

We then extend the network size to a larger scale. 214 genes are selected based on the distance between two gene vectors. If the distance is smaller than a threshold, one is removed. Based on the threshold 0.17 for the negative data set and 0.135 for the positive data set we have 24 and 23 genes in the negative and positive data sets, respectively. Figure 16.10 shows the DBN structures.

Negative DBN (whole)   **Positive DBN (whole)**

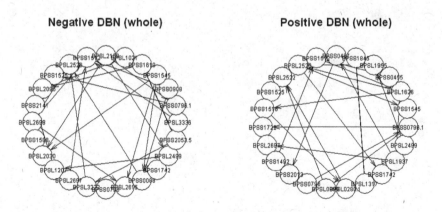

Fig. 16.10. Two DBN structures derived from 24 genes in the negative data set and 23 genes in the positive data set. The genes are selected based on a distance threshold.

Summary

This chapter has discussed the discrete Bayesian network and its applications to bioinformatics. In constructing a Bayesian network, we are required to undertake three different tasks. First, we need to learn model parameters, i.e. the posterior probabilities from each parent variable to a child variable. Second, we need to use a built network to make inferences. This involves the prediction of a certain event with a confidence. Third, we can use the network to infer how likely one "downstream" (child) variable is to be the outcome of its causative variable "upstream". Bayesian networks are a class of powerful machine-learning algorithms which can help discover the underlying gene regulatory network for understanding how a gene network responds to environmental factors and generating new hypotheses. However, Bayesian networks are known to have a vital limit in computational cost as well as network size. Currently, building a Bayesian network for data with over 1000 variables is a non-trivial task, i.e. the computational cost is not affordable. More and more new algorithms which aim to improve the efficiency and accuracy of Bayesian networks learning have been proposed. The evaluation and validation of them for general purpose Bayesian network learning is still a vital and tough task we face.

Chapter 17

S-Systems

Biochemical System Theory (BST) defines a mathematical model to study the dynamics of a biochemical system. The theory was introduced by Mike Savageau in the 1960s. Using BST to study a biochemical system does not need equations defining the exact mechanism of reactions. In a BST model, the relation between reactants and regulatory interactions is modelled by the power-law theory. The approach is referred to as S-systems or generalised mass action (GMA). There are two subjects closely related to S-system research. The first is how to learn the structure of an S-system given data. The second is how to learn system parameters given an S-system structure.

17.1 Michealis-Menten change law

Biochemical System Theory (BST) or S-system refers to a type of biochemical systems study that uses a mathematical biology approach. The model was introduced by Mike Savageau in the late 1960s [521-523]. The equations involved in BST are commonly ordinary differential equations. Each equation describes a biochemical process based on the power-law theory.

The dynamics of a biochemical system with a single component (X) can be described by a function $X(t)$ which is a function of time. An experiment often aims to find how the component changes through time. Observations are taken at several time intervals. From this, the change rate can be drawn and analysed. In mathematics, such a change rate can be defined by an ordinary differential equation as below

$$\frac{dX}{dt} = \alpha X + \beta \tag{17.1}$$

To solve this equation, an integral can be taken to generate the following equation

$$X = \frac{\beta}{\alpha}(e^{\alpha t + C} - 1) \tag{17.2}$$

If the initial state is (0, 0), the system may be plotted as in Fig. 17.1.

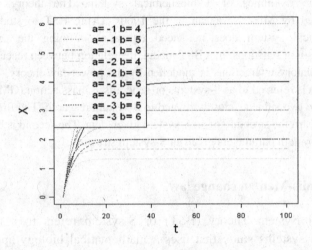

Fig. 17.1. An illustration of change rate for a single component biochemical system. The horizontal axis represents time and the vertical axis represents the quantity of the component (X).

A well-known rate change example in biochemistry is the Michealis-Menten rate law [524]. The reaction between an enzyme (E) and a substrate (S) leads to a product (P). The diagram of a simple biochemical reaction is shown in Fig. 17.2, where k_1, k_{-1} and k_2 are three reaction constants.

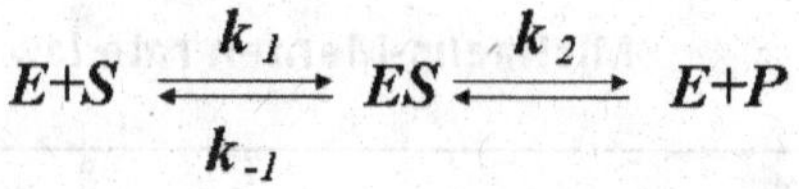

$$E+S \underset{k_{-1}}{\overset{k_1}{\rightleftharpoons}} ES \overset{k_2}{\rightleftharpoons} E+P$$

Fig. 17.2. An illustration of a simple biochemical reaction.

The system is described as

$$\frac{dS}{dt} = \dot{S} = -\frac{V_{max}S}{K_M + S} \tag{17.3}$$

where V_{max} is the maximum change rate of the substrate and K_M is called the Michealis constant. The product is described by another differential equation as below

$$\frac{dP}{dt} = \dot{P} = \frac{V_{max}S}{K_M + S} \tag{17.4}$$

The relation between them is shown in Fig. 17.3.

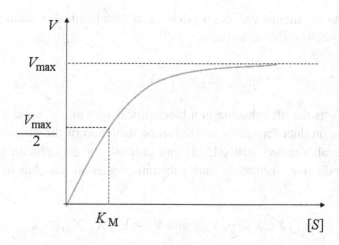

Fig. 17.3. The relation between V_{max} and K_M.

Solving the ordinary differential equations defined in equations (17.3) and (17.4) leads to the system dynamics shown in Fig. 17.4.

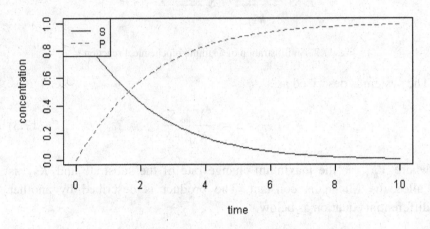

Fig. 17.4. An illustration of the system dynamics of a simple biochemical system using the Michaelis-Menten rate law. In this simulation, $V_{\max} = 2$ and $K_M = 4$.

17.2 S-system

A general mathematical description of a biochemical system of N substrates can be shown as below

$$\dot{X}_i = V_i^+ - V_i^- \tag{17.5}$$

where X_i is the ith substrate in a biochemical system., V_i^+ and V_i^- are called the product formation and substrate depletion of the ith substrate. They are also called influx [525] and outflow (or degradation [525]). Both production formation and substrate depletion are functions of substrates

$$V_i^+ = V_i^+(X_1, X_2, \cdots, X_N) \text{ and } V_i^- = V_i^-(X_1, X_2, \cdots, X_N) \tag{17.6}$$

A three-component and one-reaction system is shown in Fig. 17.5 [525]. The function of the enzyme X_3 is to convert the substrate X_1 to the product X_2. X_1 has no influx but has a degradation term. Its influx is then replaced by a constant C. X_2 has one influx term and a degradation term. The system is defined below by two ordinary differential equations

$$\dot{X}_1 = C - V_1^-(X_1, X_3)$$

$$\dot{X}_2 = V_2^+(X_1, X_3) - V_2^-(X_2)$$

(17.7)

Note that $V_1^-(X_1, X_3) = V_2^+(X_1, X_3)$.

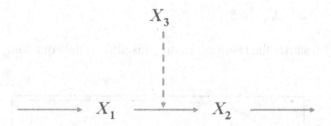

Fig. 17.5. A three-component and one-reaction biochemical system., X_1 and X_2 are substrates while X_3 is an enzyme.

The next critical question is how to represent $V_1^-(X_1, X_3)$ and $V_2^-(X_2)$ as functions of the substrates and enzyme. Based on the pioneering work of Savageau [521-523], both outflows of equation (17.7) are represented as the products of molecules, which is referred to as the power law as shown below

$$V_1^-(X_1, X_3) = aX_1^b X_3^c$$

$$V_2^-(X_2) = dX_2^e$$

(17.8)

where a, b, c, d, and e are five constants. A generalised form of a biochemical system with N substrates and N enzymes is shown below, called an S-system where "S" refers to *synergism* and *saturation* of an investigated system [525].

$$\dot{X}_i = \alpha_i \prod_{j=1}^{N+M} X_j^{\mu_{ij}} - \beta_j \prod_{j=1}^{N+M} X_j^{\nu_{ij}}, \forall i \in [1,M] \qquad (17.9)$$

where α_i is a constant of the production formation rate, β_j is a constant of the substrate depletion rate, μ_{ij} is the production formation kinetic rate, and ν_{ij} is the substrate depletion kinetic rate.

Here we show an example [525]

$$\dot{X}_1 = \alpha_1 X_2 - \beta_1 X_1^{0.5} X_3^{-1}, X_1(0) = 0.001$$
$$X_2 = 2 \qquad\qquad (17.10)$$
$$X_3 = 0.5$$

Figure 17.6 shows the result of solving this differential equation.

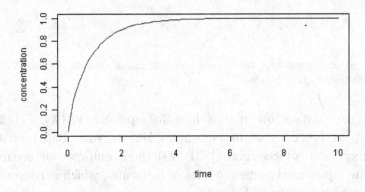

Fig. 17.6. The dynamics of the S-system described by equation (17.10), where $\alpha_1 = \beta_1 = 1$.

In Voit's book [525], three properties of an S-system are discussed. In brief, they are validity, theoretic justification, and analytic convenience. For the first one, it is shown that S-system models are consistent with real biochemical systems. In many experiments, the power-law property of an S-system is confirmed. For the second, Voit presented three arguments. One is that concentration is very close to the steady-state.

The next is the fact that the relative change of metabolite concentration is generally linear. The final one is that virtually any phenomenon can be formulated using an S-system. For the third, Voit argued that an S-system can easily be modelled numerically.

However, an S-system still has a certain distance from real applications because of the difficulty in structure identification and parameter estimation. In a real application, it is normally rare to have prior knowledge of a model structure as well as model parameters. This means that it is hard to define an exact influx function and a degradation function for each substrate. Even if the model structure can be approximately determined, a proper approach is needed to estimate all the parameters. The difficulty is even more severe when we have limited data, i.e. the time points are limited.

Various machine learning algorithms are therefore employed in various studies to address these two problems.

17.3 Simplification of an S-system

Because structure identification and parameter estimation are the most important things to do, a simplification process of an S-system is required. This will lead to several algebraic equations which can be solved by machine learning algorithms. The simplification is to make a discrete version of an S-system. Suppose we have collected a data set comprising T sets of observations for N substrates and M enzymes. This process is also called decoupling [526, 527]. The process is to replace the left-hand side of the differential equations of an S-system by gradients which are estimated in several ways described in the next section. If gradients at T time points are calculated, $(T \times N)$ nonlinear equations can be defined as below

$$S_i(t) = \alpha_i \prod_{j=1}^{N+M} X_j^{\mu_{ij}}(t) - \beta_j \prod_{j=1}^{N+M} X_j^{v_{ij}}(t) + \varepsilon_i(t) \qquad (17.11)$$

where $\varepsilon_i(t)$ is the error term meaning that the gradients calculated may have some deviations from the outputs of the true models ($y_i(t)$)

$$y_i(t) = \alpha_i \prod_{j=1}^{N+M} X_j^{\mu_{ij}}(t) - \beta_j \prod_{j=1}^{N+M} X_j^{\nu_{ij}}(t) \qquad (17.12)$$

An error function is defined as below

$$\sum_{t=1}^{T}\sum_{i=1}^{M}(\varepsilon_i(t))^2 = \sum_{t=1}^{T}\sum_{i=1}^{M}(S_i(t) - y_i(t))^2 \qquad (17.13)$$

Through minimising this error function, model parameters can be estimated.

17.4 Approaches for structure identification and parameter estimation

In order to identify a proper S-system structure and estimate its parameters, we are required to use a specific approach if we have a specific assumption. Three approaches have been used for structure identification and parameter estimation. They are neural network approach, evolutionary computation approach, and steady-state approach.

17.4.1 *Neural network approach*

With the neural network approach, the gradients are estimated using neural network algorithms [526-528]. A neural network model is built for estimating gradients. Its inputs are the times and its outputs are the gradients. The observed gradients at t_j for X_i is estimated from experimental data by

$$\dot{X}_i(t_j) = \frac{X_i^{t_j} - X_i^{t_{j-1}}}{t_j - t_{j-1}} \qquad (17.14)$$

Figure 17.7 shows an example of using neural network to estimate gradients. The raw data are the same as in the simple change rate example shown in Fig. 17.1 where $\alpha = -3$ and $\beta = 2$. Two hidden neurons are used. It can be seen that neural network can well estimate the gradients with a small error.

After gradients have been estimated, the ordinary differential equations become nonlinear algebraic equations for which Newton-Raphson numerical approximation algorithm [529] can be used to estimate model parameters [525].

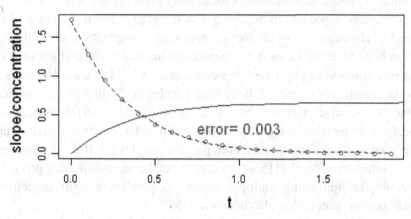

Neural network for estimating slopes

Fig. 17.7. An illustration of estimating gradients using neural network. The circles are the estimated gradients from raw data. The solid line represents the concentration. The dashed line represents the predicted gradients using neural network.

17.4.2 *Simulated annealing approach*

With simulated annealing approach, the main objective is to determine the model parameters [530]. The same optimisation principle is applied as when using neural network algorithm, i.e. minimising the error between model outputs and experimental data. However, unlike neural network approach mentioned above, gradients are not estimated. Instead, the first order approximation is used as below

$$X_i(t_j) \approx X_i(t_j) + \dot{X}_i(t_{j-1})\Delta t_j \qquad (17.15)$$

where $\dot{X}_i(t_{j-1})$ is evaluated at time t_{j-1} and Δt_j is the time interval from time t_{j-1} to time t_j. Simulated annealing (SA) is an global optimisation approach used in machine learning [531].

17.4.3 *Evolutionary computation approach*

Similar to simulated annealing, evolutionary computation is also a global optimisation approach in machine learning [532, 533]. In using the genetic algorithm, one of the evolutionary computation algorithms, Kikuchi et. al. code the model parameters using real codes and search for the best solution through evolving candidates [534]. In Edwards's work, model parameters are coded using binary codes [535]. In Kimura's work, gene time-course data are estimated using spline interpolation or a local linear regression technique. The estimated gene time-course data are then used in an optimisation process employing the cooperative coevolutionary algorithm [536]. In order to make an evolutionary process suitable for optimising multiple criteria in parallel, a multi-objective optimisation strategy has also been used [537].

17.5 Steady-state analysis of an S-system

Rather than analysing dynamical properties and functions, many experiments focus on steady-state analysis as it contains important biochemical patterns [525]. In this situation, we assume $\dot{X}_i \equiv 0$. From this, equation (17.9) is re-written as

$$\alpha_i \prod_{j=1}^{M} X_j^{\mu_{ij}} = \beta_j \prod_{j=1}^{M} X_j^{\nu_{ij}}, \forall i \in [1, M] \qquad (17.16)$$

Applying logarithm to the above equation leads to

$$\ln \alpha_i + \sum_{j=1}^{M} \mu_{ij} \ln X_j = \ln \beta_j + \sum_{j=1}^{M} v_{ij} \ln X_j, \forall i \in [1,M] \quad (17.17)$$

Given all the model parameters, we can analyse the relation between substrates and enzymes at the steady-state. Equation (17.17) can be re-written as

$$\sum_{j=1}^{M} (\mu_{ij} - v_{ij}) \ln X_j = \ln \frac{\beta_j}{\alpha_i}, \forall i \in [1,M] \quad (17.18)$$

For instance, suppose we have a biochemical system defined as below

$$\dot{X}_1 = 0.5 X_2^{-2} - 0.2 X_1$$
$$\dot{X}_2 = 0.2 X_1 - 0.5 X_2^{0.5} \quad (17.19)$$

Its dynamics are shown in Fig. 17.8, where we assume the initial states is at (1, 1).

The steady-state analysis is shown as below

$$\ln X_1 + 2 \ln X_2 = \ln \frac{5}{2}$$
$$\ln X_1 - \frac{1}{2} \ln X_2 = \ln \frac{5}{2} \quad (17.20)$$

The equations give

$$X_1 = 2.5$$
$$X_2 = 1 \cdot \quad (17.21)$$

From Fig. 17.8, it can be seen that the steady-state calculation is correct. Figure 17.9 shows the trace of the system which approaches the steady-state at the point (2.5, 1) through time. It can be seen that the curve reaches the final point at (2, 5) from the initial sates at (1, 1).

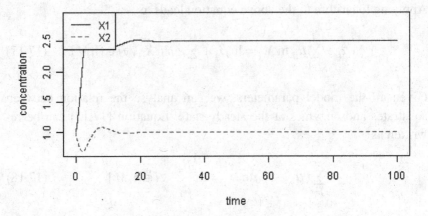

Fig. 17.8. A biochemical system to show steady-state study.

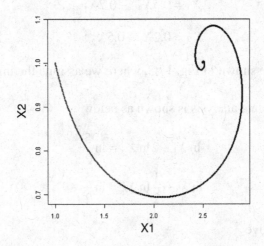

Fig. 17.9. The trace of the system. The horizontal axis represents X_1 while the vertical axis represents X_2.

Feedback is a common phenomenon in biochemical systems. Here we consider the application of the S-system to a biochemical system with four molecules. The system is shown in Fig. 17.10 and is quantified as below

$$\dot{X}_1 = 3X_3^{1.8}X_2^{-5} - X_1^{1.2}$$

$$\dot{X}_2 = X_1^{1.2} - 3X_2^{0.8}X_4^{-5}$$

(17.22)

The initial state is

$$X_1 = 1$$
$$X_2 = 1$$
$$X_3 = 10$$
$$X_4 = 10$$

(17.23)

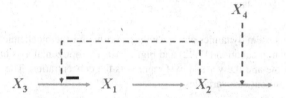

Fig. 17.10. A biochemical system with a feedback. X_3 and X_4 are the inputs to this system while X_2 has a feedback on X_1. The solid lines represent the reactions while the broken lines represent enzymatic activities.

Figure 17.11 shows the system dynamics (the left panel) and the trace through time (the right panel).

Another even more complicated feedback system can be considered as below. Suppose the feedback is made through a non-linear function, say a sigmoid function described as below

$$\rho(X) = \frac{1}{1 + \exp(-X)}$$

(17.24)

We can formulate a complicated feedback system with the same initial state defined in equation (17.23) as below

$$\dot{X}_1 = 3X_3^1 X_2^{-2} - \rho(X_1)$$

$$\dot{X}_2 = \rho(X_1) - 3X_2^1 X_4^{-2}$$

(17.25)

The simulation result is shown in Fig. 17.12.

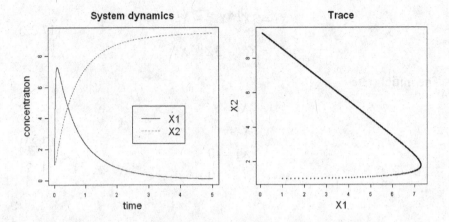

Fig. 17.11. The system dynamics and trace through time for the biochemical system with a feedback shown in equation (17.22) and Fig. 17.10. The horizontal axis in the left graph represents the time and the vertical axis represents the concentration. The horizontal and vertical axes represent two variables.

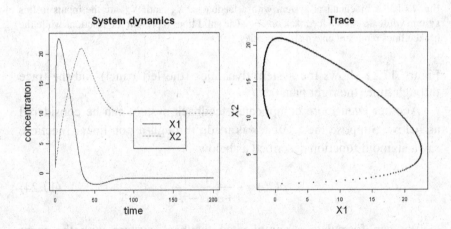

Fig. 17.12. The simulation of a system with sigmoid feedback defined in equation (17.25). The horizontal axis in the left graph represents the time and the vertical axis represents the concentration. The horizontal and vertical axes represent two variables.

The above two cases show that the S-system can be well used for analysing complicated biochemical systems with feedbacks.

17.6 Sensitivity of an S-system

In the steady-state, it is also required to analyse how a variable varies with changes in other variables. This is called a sensitivity analysis which is an analysis of how a biochemical system is sensitive to a small perturbation at the steady-state [525]. The sensitivity analysis is done through the calculation of gains. In Voit's book [525], five types of gains are introduced. Only the log gain of the steady-state concentration between dependent variables and independent variables is discussed here. For more details, chapter 7 of Voit's book is recommended.

A log gain between two variables is defined as

$$S_{ij} = \frac{\log X_i}{\log X_j} \tag{17.26}$$

where X_i and X_j are the two variables. In most situations, we are interested in analysing the relation between a dependent variable and an independent variable. Therefore, X_i and X_j should be a dependent and an independent variable, respectively. S_{ij} measures how sensitive $\log X_i$ is to a small perturbation in $\log X_j$.

We use the system defined in equation (17.22) for the following discussion. The steady-state of the system is defined as below

$$\log 3 + 1.8 y_3 - 5 y_2 = 1.2 y_1$$
$$1.2 y_1 = \log 3 + 0.8 y_2 - 5 y_4 \tag{17.27}$$

where $y_i = \log X_i$. Here y_3 and y_4 are independent variables while y_1 and y_2 are dependent variables. From the above equations, we can see that

$$y_1 = 0.91551 + 0.2069 y_3 - 3.592 y_4$$
$$y_2 = -0.8789 + 1.55172 y_3 + 4.3104 y_4 \tag{17.28}$$

The log gain between X_1 and X_3 is 0.2069 while the log gain between X_1 and X_4 is -3.592. The other two log gains can be seen in the above equation. Figure 17.13 shows the sensitivity maps (log gain maps) of two dependent variables on the two independent variables.

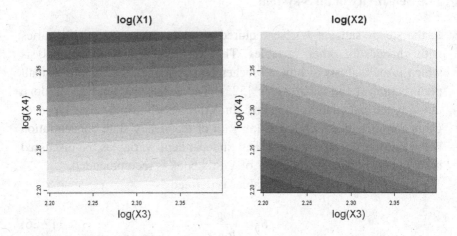

Fig. 17.13. The sensitivity maps of two dependent variables on two independent variables. The left panel is the log gain map for X_1 while the right panel is the log gain map for X_2.

Summary

This chapter has discussed the S-system approach which studies the dynamics of a biochemical system. Because of the complexity and nonlinearity, two issues (structure identification and model parameter estimation) are still challenging. In most experiments, it is difficult to know most structures, pathways or network structures in advance. This makes the modelling process difficult because there are too many different structures for selection. Researchers have done tremendous work for various applications. However, we still have a long way to go for the S-system to be applicable for large systems with required accuracy and efficiency. Nevertheless, the S-system approach is a vital contribution to modern systems biology research. Its advantages are obvious in exploring underlying biochemical patterns and generating new hypotheses for further experiments. In addition to structure identification and parameter estimation, steady-state analysis and log gain sensitivity analysis between variables have also been discussed.

Chapter 18

Future Directions

Although bioinformatics has been well-developed for a few decades with the enhancement of machine learning approaches, there are still some challenges. Many of these result from the gap between fast technology development and slow software development. For instance, the next-generation sequencing technology can speed up sequencing, but leaves a problem of fragment errors and sequence assembly. This is because fragment assembly algorithms have to be adapted to handle very short fragments. This kind of issue is related to computing skills and parallel computing concept and will not be discussed in this book. However, in using machine learning algorithms for modelling biological data, we also have many challenges. The most typical ones are the multi-source data and data size. Most machine learning algorithms deal with single-source data. Systems biology research, which is one of the major research themes this century, aims to analyse a biological system systematically. Multiple data sources certainly need consideration. The data size is always a huge challenge in bioinformatics. Handling hundreds of thousands or millions of DNA fragments for species diversity study is an example. Even modern computers still face challenges in modelling entire data and some specific data treatment has to be introduced. One such treatment is to lose possible resolution, for instance reducing 64 bits to 32 bits for representing numerical data. Gene expression data analysis for cancer diagnosis and drug development is also a challenge in using machine learning approaches for bioinformatics. One of the challenges in this area is to deal with heterogeneity in data. Most existing algorithms assume homogeneity in data and have missed one important issue in cancer diagnosis that each patient may develop a distinct biochemical system to develop cancer disease. Ignoring such heterogeneity may lead

269

to the developed drug being hardly applicable to a large variety of patients. How to handle this challenge has been one of the most important subjects of the century. In this chapter we discuss these typical challenges in bioinformatics.

18.1 Multi-source data

Compared with genomics and other omics, metabolite data contain much more information for biological discovery [538]. This is related to the fact that the metabolome is called life indicator across most living organisms. It is known that there are about 3 billion base pairs in the human genome, but there are only about 3,000 to 100,000 metabolites. The difficulty is that most metabolites are unknown and the discovery of new metabolites is still increasing rapidly. This means that the knowledge of existing metabolite pathways and networks is constantly being updated. It then lays a difficulty in using the existing databases for inference.

Although metabolite separation can be completed by existing chromatography technologies such as NMR, GC-MS, LC-MS, and LC-MS/MS, identification of each metabolite is a difficult task. There is still no technology capable of separating ions perfectly. Multi-dimensional technique has been used for better separation [539-544]. This then casts a problem with most existing machine learning approaches which are aimed at analysing one-dimensional metabolite data.

In comparing NMR and MS data, it is found that there is a small overlap of the metabolites found and it is concluded that both techniques only cover a part of the whole metabolome and neither of them is able to provide the whole picture [545].

Both NMR and MS have a problem of sensitivity when producing metabolite profiles. This means that analysing metabolite profile data with variable sensitivity using machine learning approaches is still a typical and difficult problem.

Even some software, such as one provided by Agilent Technologies, can predict metabolite formulae for experimental metabolites. However, each formula may be mapped to multiple compounds. Inferring differential metabolite pathways activated by experiments is then a difficult task. Although daughter ions information can be used, paring parent/daughter ions needs some carefully prepared learning processes and the relationship between daughter ion spectrums and log P values which are unique to compounds is another difficult regression problem where the issue of equipment variation must be well-measured. This is largely related to equipment settings. There are also many other issues such as solute and environmental factors which can contribute to possible variations making spectral separation a difficult task at present, needing sophisticated bioinformatics techniques.

The second issue in multi-source data is very common in systems biology, i.e. how to combine expression data and metabolite data for pathway and network analysis. It is known that mapping a gene to a pathway has a higher accuracy compared to mapping a metabolite to a compound, then a pathway. This is because of two reasons. Compared with genomics, metabolomics still has a short history and it lacks well-developed databases of annotated compound information. However, compared to gene chip data, metabolites can be used to explore phenotypic patterns more extensively. In systems biology research, a challenge is how to combine both gene expression data and metabolite data together for analysis. For instance, a system called springScape has been developed for combining several relevant data sources (protein-protein interaction information and Gene Ontology (GO) annotations) for clustering gene expression data [546]. MetaLook software was developed for visualising and analysing marine ecological genomic and metagenomic data (i.e. of environmental relevance) with respect to habitat parameters. The software is able to map relevant genomic information onto a world map [547]. In conjunction with a phylogenetic tree model, a visualisation tool was implemented to analyse the helix-loop-helix transcription factor interaction network. From this, the tool allows the user to clarify the context of network hubs and interaction clusters [548]. A 3-D visualisation tool was designed for the analysis of functional linkage between genes in large data sets [549]. A web-based visualisation tool was implemented for delivering an intuitive, interactive environment for constructing ontology-based queries against the GO

Database [550]. A web-based visualisation toolkit called VariVis was built for the analyses of mutation data related to diseases, where multiple databases can be combined [551]. Recently, with democratisation of mass spectrometry and the realisation that small molecules were central to determining phenotypes, visualisation of metabolite data has emerged as an important objective. Metabolite data has, for instance, first been partitioned; regression analysis has then been used to analyse and visualise the relationships between metabolite and gene data [552]. Like gene expression data, metabolite data also have the problem that the number of samples is much smaller than the number of variables (metabolites). Because of this, PCA and independent component analysis are used for data dimension reduction and visualisation [553]. The visualisation is based on the first two (principal/independent) components.

18.2 Gene regulatory network construction

Constructing gene regulatory networks is one of the most important *in silico* exercises in systems biology and bioinformatics for exploring complexity in data and generating hypotheses for further scientific verification which then generates new evidence and new data for further *in silico* study. The challenge for bioinformatics and machine learning approaches is to be able to model given data with the desired accuracy and efficiency, and importantly consistency with biology. Inferring gene regulatory networks uses two different approaches:, bottom-up and top-down approaches [554]. As mentioned in chapter 16, gene regulatory network construction has two tasks to do. One is structure identification and the other is model parameter estimation. The bottom-up approach and top-down approach actually deal with these two concepts separately. The bottom-up approach starts from a detailed mathematical description of a biochemical process for model parameter estimation. The basic technique used for this is called the S-system described in chapter 17. With the S-system technique, differential equations are defined according to molecular interactions as an *a priori* leaving model parameters estimated from observations. The top-down approach constructs gene regulatory networks based on data, which is also called the data-driven approach. Bayesian networks described in chapter 15 are one of such techniques. Compared with bottom-up approach, the top-down approach is more appropriate because it does not need *a prior* knowledge of biochemical process in a system which is to be modelled. However, as with most structure identification in machine learning, the top-down approach has its own limitation, i.e. uncertainty. The uncertainty mainly results from the quality of single source data. It is then recommended to use multiple sources for gene regulatory network construction [554]. Combining gene expression, genomic sequences and protein-DNA interaction data is such a practice [555]. The sources which can be considered for gene regulatory network construction are gene expression, transcription factor binding sites, genomic sequences, and chromatin immunoprecipitation (ChIP) [554]. With multiple data sources, typical issues are the quality, dimensions, and format of heterogeneous data. All these three issues are very practical. Heterogonous data quality means that we have to consider heterogonous models; each has a different technique to handle the uncertainty in data. Heterogonous data dimension means that some specific data treatment must be considered to avoid any possible extra introduction of uncertainty. For instance, increasing the dimensionality of one data source may introduce noise so that we have to consider random sampling which certainly increases data size. Reducing data dimensionality may lose some important information contained in a data set. Heterogonous data format needs a careful consideration of data integration. For instance, gene expression is numeric while genomic sequence is non-numeric. Integrating them into one system for gene regulatory network construction needs specific design of machine learning algorithms so as to handle possible uncertainty. For a detailed review about this challenge, readers can refer to review papers [554, 555] [556].

Differential equation parameters of an S-system can be estimated using artificial neural networks, evolutionary computation, and simulated annealing as described in Chapter 17. The network size is still mediate in most recent applications. First, fully simulating a large differential equations system is time consuming. Second, setting up a large S-system for a biochemical system is still challenging. Third, most biochemical systems have limited time points of observations. This means that converting differential equations to difference equations and then using machine learning approaches to estimate differential equation parameters can generate imprecise models.

Bayesian networks technique described in chapter 16 also has a limitation in computing cost. It is difficult to make a Bayesian network with over 100 nodes. This is because the searching space of network structures is increased significantly when the number of nodes is increased [557, 558]. Yet another problem of Bayesian networks approach is that it is extremely difficult to learn Bayesian network parameters if training data is incomplete or sparse [559]. However, incompleteness and sparseness often occur in biological experiments.

Developing novel techniques is extremely desirable when the biochemical network size is constantly increasing.

18.3 Building models using incomplete data

Described in chapter 15, peptide classification is based on the assumption that the data collected is complete. In fact, this is hardly true. New functional sites are continually being discovered. For other applications that use machine learning algorithms, incompleteness only means that a small proportion of the data is missing or wrongly labelled. This may not be true in peptide classification. In most biological experiments, only functional sites are of interest. This means that nearly all the "sites" collected in databases are only functional sites. For instance, they can be phosphorylation sites, acetylation sites, methylation sites, protease cleavage sites, binding sites, active sites, etc. In a machine-learning or classification process, they are treated as positive data. In order to make a model capable of predicting novel data, it must be trained to discriminate between positive and negative data. For instance, enabling a predictive model capable of predicting phosphorylation sites, we must teach the model how to recognise what a phosphorylation site looks like and what a non-phosphorylation site looks like. However, there may never be any experimentally annotated or verified negative data (sites). A practical or popular approach is to use non-annotated sites as negative data. For instance, methylation happens to lysine or proline amino acids. A residue occupied by an amino acid other than these two amino acids will not have any possibility of being methylated; the possibility is only there for lysine and proline. Whenever we find a lysine or a proline, we need to predict if it is involved in methylation. Not all lysines or prolines are involved in methylation. This is because the substrate pattern or specificity determines whether methylation occurs. We normally assume all non-annotated lysine and proline residues to be negative data. A problem arises! Without experimental verification, by what justification can we definitely say a non-annotated lysine or proline will never involve methylation? The answer is none. This means that a model built this way will not be able to predict a methylation site whose pattern has been included as negative data in a training data set. The diversity of experimentally annotated functional sites is an important causative agent of many current predictors which do not have high prediction accuracy for novel data. The challenge for machine learning is how to enlarge its error-tolerance rate or how to evaluate the possibility that such a model will miss some novel functional sites.

18.4 Biomarker detection from gene expression data

Identifying or ranking differentially expressed genes is closely related to medicine and drug development for human health and other related issues in fighting plant disease and improving animal health. A highly ranked gene is often called a biomarker. It can be used for disease diagnosis and drug development [560-563]. Conventional approaches still largely rely on T test or some simple approach. Although feature selection approach and machine learning approaches [564, 565] have been used in some stages, their usability is still in question.

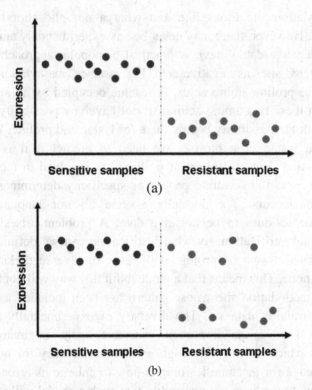

Fig. 18.1. (a) Mock expression pattern from a gene which is detected by t-test as a perfect marker when it is differentially expressed in one class compared to another class. In this model, the expression level of all sensitive samples is higher than for all resistant samples. (b) Mock expression pattern from a gene likely to be involved in a heterogeneous characteristic of a cancer phenotype; in one group, resistant, samples do not show uniform down-expression for a gene. Blue and red circles illustrate the expression level of sensitive and resistant samples, respectively.

Conventional methods commonly make an inappropriate assumption that a good biomarker shows a perfect shift in expression value of all samples, such as the significance analysis of microarrays which is a derivation of the T test, (SAM) [566] and CyberT [567], correlation analysis or fold change analysis. Such a naïve assumption can hardly be true. Taking human disease as an example it is known that each person is a system, a biochemical system in which the biochemical process can be very different from that of other people who have caught the same disease [568]. Multiple unrelated mechanisms, which are unknown to the experimenter, might independently result in resistance to a drug such as absence of signalling from the target, presence of compensatory signals reducing dependency on the target or presence of mechanisms preventing drug activity. It is likely that any one gene involved in only one such mechanism will only show differential expression in the subset of resistant cell lines driven by that particular mechanism. Figure 18.1 shows such a possibility.

A good example is ERBB2 (HER2) which is hyper-expressed in only 15–20% of breast tumours compared with normal breast tissues [569]. Such a gene may, therefore, be ranked low by t-test P-value; however, the gene is highly relevant and valuable as a resistance marker.

A number of approaches have been proposed for addressing this in different settings. Outlier detection methods, such as Cancer Outlier Profile Analysis (COPA) [570] "outlier-sum" [571], outlier robust t-statistics (ORT) (Liu and Wu, 2007) and maximum ordered subset T-statistics (MOST) [572] aim to identify subsets of samples that show marked differential expression from the majority of samples. More recently, van Wieringen et al. [573] designed a statistical test, named PDGEtest, to detect partial shifts in expression values of sub-samples. These methods have been demonstrated to outperform typical univariate statistical methods in specific settings, for example identification of expression from amplified genes in prostate cancer [570]. These recently created methods demonstrate the potential to enhance comprehensive analysis results and power over traditional methods of differential gene-expression detection in heterogeneous samples, suggesting application of these methods for gene selection could result in better overlap between different gene lists. The poor reproducibility in results from the different algorithms, however, suggests none may offer a complete solution. They are mainly reliant on detection of outliers and as such only select for genes displaying clear bimodal expression patterns. However, genes may also show non-normal distribution patterns. Furthermore, gene scoring and ranking by such methods typically reflects the size of the outlying groups without considering exclusiveness of this group to a particular phenotype of interest; clarity and size of differential expression from the sub-group, and the proportion of samples of that phenotype that the outlying sub-group represents.

A promising technique for coping with these challenges is to consider heterogeneous characteristics of samples and rank genes based on the expression pattern of sub-samples rather than on the average expression levels of phenotypic groups. To optimize reproducibility, genes ranked high should exhibit high or low expression in only a subset of samples which are treated as outliers to cover some new criteria such as exclusiveness, clearness, fold change, and proportion of differentially expressed samples.

Summary

This chapter has discussed a few typical challenges in using machine learning approaches for modelling biological data. They are mainly multi-source data, data size, and biomarker identification. It must be noted that aside from these three typical challenges, many others can be even more important such as metagenomics, mass spectrometry-based proteomics, compound identification, protein identification, etc. However, this book is mainly tailored for using machine learning approaches for bioinformatics, where we need to focus on how to improve the existing machine learning algorithms to solve the problems we currently have.

References

[1] T.K. Attwood, Parry-Smith, D.J., *Introduction to bioinformatics*. Essex: Addison Wesley Longman Ltd, 1999.

[2] A.D. Baxevanis, Ouellette, B.F.F., *Bioinformatics, A Practical Guide to the Analysis of Genes and Proteins* vol. 39. New York: John Wiley & Sons, Inc, 1998.

[3] P.G. Higgs, Attwood, T.K., *Bioinformatics and Molecular Evolution*. MA, USA: Blackwell Science Ltd, 2005.

[4] P. Baldi, Brunak, S., *Bioinformatics, The Machine Learning Approach*. London: The MIT Press, 2000.

[5] D.W. Mount, *Bioinformatics, Sequence and Genome Analysis*. New York: Cold Spring Harbor Laboratory Press, 2001.

[6] J. Augen, Bioinformatics in the Post-Genomic Era, Genome, Transcriptome, Proteome, and Information-Based Medicine. Boston: Addison-Wesley, 2005.

[7] I. Eidhammer, Jonassen, I., Taylor, W.R., *Protein Bioinformatics, An Algorithm Approach to Sequence and Structure Analysis*. West Sussex: John Wiley & Sons, Ltd, 2004.

[8] G.J. Morgan, "Emile Zuckerkandl, Linus Pauling, and the Molecular Evolutionary Clock, 1959-1965," *Journal of the History of Biology*, vol. 31, pp. 155-178, 1998.

[9] E. Zuckerkandl, "On the Molecular Evolutionary Clock," *Journal of Molecular Evolution*, vol. 26, pp. 34-46, 1987.

[10] S.F. Altschul, Gish, W., Miller, W., Myers, E.W., Lipman, D.J., "Basic local alignment search tool," *J Mol Biol*, vol. 215, pp. 403-410, 1990.

[11] S. Needleman, Wunsch, C., "A general method applicable to the search for similarities in the amino acid sequence of two proteins," *J Mol Biol*, vol. 48, pp. 443-53, 1970.

[12] P.H. Sellers, "On the theory and computation of evolutionary distances," *SIAM J Appl Math*, vol. 26, pp. 787-93, 1974.

[13] T.F. Smith, Waterman, M.S., "Identification of Common Molecular Subsequences," *Journal of Molecular Biology*, vol. 147, pp. 195-197, 1981.

[14] W.J. Wilbur, Lipman, D.J., "Rapid similarity searches of nucleic acid and protein data banks," *Proc. Natl. Acad. Sci*, vol. 80, pp. 726-730, 1983.

[15] M.O.S. Dayhoff, R.M., "Atlas of Protein Sequence and Structure," *Nat. Biomed. Res. Found,* vol. 3, pp. 353-358, 1978.

[16] R.M. Schwartz, Dayhoff, M.O., "Protein and Nucleic Acid Sequence Data and Phylogeny," *Science,* vol. 205, pp. 1036-1039, 1979.

[17] R.V. Eck, Dayhoff, M.O., "Evolution of the structure of ferredoxin based on living relics of primitive amino acid sequences," *Science,* vol. 152, pp. 363-366, 1966.

[18] R.M. Schwartz, Dayhoff, M.O., "The Point Mutation Process in Proteins," in *Proc. 2nd ISSOL and 5th ICOL Meeting,* Japan, 1977, pp. 457-469.

[19] S.B. Needleman, Wunsch, C.D., "A general method applicable to the search for similarities in the amino acid sequence of two proteins," *J Mol Biol,* vol. 48, pp. 443-53, 1970.

[20] R. Staden, "Sequence data handling by computer," *Nucleic Acids Res,* vol. 4, pp. 4037-51, 1977.

[21] R.F. Doolittle, "Similar amino acid sequences: chance or common ancestry?," *Science,* vol. 214, pp. 149-59, 1981.

[22] D.J. Lipman, Pearson, W.R., "Rapid and sensitive protein similarity searches," *Science. 1985 Mar 22;(4693):,* vol. 227, pp. 1435-41, 1985.

[23] S.F. Altschul, Gish, W., Miller, W., Myers, E.W., Lipman, D.J., "Basic local alignment search tool," *J. Mol. Biol,* vol. 215, pp. 403-410, 1990.

[24] H. Pearson, "Genetics: what is a gene?," *Nature* vol. 441, pp. 398-401, 2006.

[25] E. Pennisi, "DNA study forces rethink of what it means to be a gene," *Science* vol. 316, pp. 1556-1557, 2007.

[26] M.B. Gerstein, Bruce, C., Rozowsky, J.S., Zheng, D., Du, J., Korbel, J.O., Emanuelsson, O., Zhang, Z.D., Weissman, S., Snyder, M., "What is a gene, post-ENCODE? History and updated definition," *Genome Research,* vol. 17, pp. 669-681, 2007.

[27] R.M. Steinman, Moberg, C.L., "A triple tribute to the experiment that transformed biology," *J. Exp. Med,* vol. 179, pp. 379-84, 1994.

[28] E.M. Southern, "Detection of specific sequences among DNA fragments separated by gel electrophoresis," *J Mol Biol,* vol. 98, pp. 503-517, 1975.

[29] D.A. Kulesh, Clive, D.R., Zarlenga, D.S., Greene, J.J., "Identification of interferon-modulated proliferation-related cDNA sequences," *Proc Natl Acad Sci USA,* vol. 84, pp. 8453-8457, 1987.

[30] M.B. Eisen, Spellman, P.T., Brown, P.O., Botstein, D., "Cluster analysis and display of genome-wide expression patterns," *Proc Natl Acad Sci USA,* vol. 95, pp. 14863-8, 1998.

[31] C. Nakada, Matsuura, K., Tsukamoto, Y., Tanigawa, M., Yoshimoto, T., Narimatsu, T., and L.T. Nguyen, Hijiya, N., Uchida, T., Sato, F., Mimata, H., Seto, M., Moriyama, M., "Genome-wide microRNA expression profiling in renal cell carcinoma: significant down-regulation of miR-141 and miR-200c," *J Pathol,* vol. 216, pp. 418-27, 2008.

[32] A.L. Aspler, Bolshin, C., Vernon, S.D., Broderick, G., "Evidence of inflammatory immune signaling in chronic fatigue syndrome: A pilot study of gene expression in peripheral blood," *Behav Brain Funct,* vol. 4, p. 44, 2008.

[33] M. Iida, Iizuka, N., Sakaida, I., Moribe, T., Fujita, N., Miura, T., Tamatsukuri, S., H. Ishitsuka, Uchida, K., Terai, S., Tokuhisa, Y., Sakamoto, K., Tamesa, T., Miyamoto, T., and Y. Hamamoto, Oka, M., "Relation between serum levels of cell-free DNA and inflammation status in hepatitis C virus-related hepatocellular carcinoma," *Oncol Rep,* vol. 20, pp. 761-5, 2008.

[34] B. Gur-Dedeoglu, Konu, O., Kir, S., Ozturk, A.R., Bozkurt, B., Ergul, G., Yulug, I.G., "A resampling-based meta-analysis for detection of differential gene expression in breast cancer," *BMC Cancer,* vol. 396, p. 396, 2008.

[35] M. Thomassen, Tan, Q., Kruse, T.A., "Gene expression meta-analysis identifies metastatic pathways and transcription factors in breast cancer," *BMC Cancer,* vol. 8, p. 394, 2008.

[36] R.N. Jorissen, Lipton, L., Gibbs, P., Chapman, M., Desai, J., Jones, I.T., Yeatman, T.J., East, P, Tomlinson, I.P., Verspaget, H.W., Aaltonen, L.A., Kruhaffer, M., Orntoft, T.F., Andersen, C.L., Sieber, O.M., "DNA copy-number alterations underlie gene expression differences between microsatellite stable and unstable colorectal cancers," vol. 14, pp. 8061-9, 2008.

[37] G.J. Weiss, Kingsley, C., "Pathway targets to explore in the treatment of non-small cell lung cancer," *J Thorac Oncol,* vol. 3, pp. 1342-52, 2008.

[38] R.M. Shai, Reichardt, J.K., Chen, T.C., "Pharmacogenomics of brain cancer and personalized medicine in malignant gliomas," *Future Oncol,* vol. 4, pp. 525-34, 2008.

[39] T. Bonome, Levine, D.A., Shih, J., Randonovich, M., Pise-Masison, C.A., Bogomolniy, F., and L. Ozbun, Brady, J., Barrett, J.C., Boyd, J., Birrer, M.J., "A gene signature predicting for survival in suboptimally debulked patients with ovarian cancer," *Cancer Res,* vol. 68, pp. 5478-86, 2008.

[40] L. Von Bertalanffy, Problems of Life, An Evaluation of Modern Biological and Scientific Thought: Harper & Brothers, 1960.

[41] M.D. Mesarovic, *Systems Theory and Biology*: Springer-Verlag, 1968.

[42] A. Agrawal, "New institute to study systems biology," *Nat Biotechnol. 1999 Aug;17(8):,* vol. 17, pp. 743-4, 1999.

[43] A. Tiselius, "A new apparatus for electrophoretic analysis of colloidal mixtures," *Trans. Faraday Soc,* vol. 33, pp. 524-531, 1937.

[44] L. Pauling, Corey, R.B., and Branson, H.R., "Two Hydrogen-Bonded Helical Configurations of the Polypeptide Chain," *Proc. Natl. Acad. Sci. USA,* vol. 37, pp. 205-211, 1951.

[45] L. Pauling, Corey, R.B., "Configurations of polypeptide chains with favored orientations around single bonds: Two new pleated sheets," *Proc. Natl. Acad. Sci. USA,* vol. 37, pp. 729-740, 1951.

[46] J.D. Watson, Crick, F.H.C., "A Structure for Deoxyribose Nucleic Acid," *Nature,* vol. 171, pp. 737-738, 1953.

[47] F. Sanger, Thompson, E.O. and Kitai, R., "The amide groups of insulin," *Biochem J,* vol. 59, pp. 509-518, 1955.

[48] S.N. Cohen, Boyer, H.W., "Process for producing biologically functional molecular chimeras. 1979.," *Biotechnology,* vol. 24, pp. 546-555, 1992.

[49] F. Sanger, Coulson, A.R., "A rapid method for determining sequences in DNA by primed synthesis with DNA polymerase," *J Mol Biol,* vol. 94, pp. 441-448, 1975.

[50] F. Sanger, Nicklen, S. and Coulson, A.R., "DNA sequencing with chain-terminating inhibitors," *Proc Natl Acad Sci U S A.,* vol. 74, pp. 5463-5467, 1977.

[51] K.H. Weisgraber, Troxler, R.F., Rall, S.C., Mahley, R.W., "Comparison of the human, canine and swine E apoproteins," *Biochem Biophys Res Commun,* vol. 95, pp. 374-377, 1980.

[52] J. Aerssens, Armstrong, M., Gilissen, R., Cohen, N., "The human genome: an introduction," *Oncologist,* vol. 6, pp. 100-109, 2001.

[53] D.A. Benson, Boguski, M.S., Lipman, D.J., Ostell, J., and Ouellette, B.F., "GenBank," *Nucl. Acids Res,* vol. 26, pp. 1-8, 1998.

[54] D.A. Benson, Karsch-Mizrachi, I., Lipman, D.J., Ostell, J., Wheeler, D.L., "GenBank," *Nucleic Acids Res,* vol. 36, pp. D25-D30, 2008.

[55] A. Cochrane G., R., Bonfield, J., Bower, L., Demiralp, F., Faruque, N., Gibson, R., Hoad, G., Hubbard, T., Hunter, C., Jang, M., Juhos, S., Leinonen, R., Leonard, S., Lin, Q., Lopez, R., Lorenc, D., McWilliam, H., Mukherjee, G., Plaister, S., Radhakrishnan, R., Robinson, S., Sobhany, S., Hoopen, P.T., Vaughan, R., Zalunin, V., Birney, E., "Petabyte-scale innovations at the European Nucleotide Archive.," *Nucleic Acids Res,* vol. 37, p. D19, 2009.

[56] M.P. Lefranc, Giudicelli, V., Kaas, Q., Duprat, E., Jabado-Michaloud, J., Scaviner, D., Ginestoux, C., Clément, O., Chaume, D., Lefranc, G., "IMGT, the international ImMunoGeneTics information system.," *Nucleic Acids Res,* vol. 33, pp. D593-D597, 2005.

[57] S. Karlin, Altschul, S.F., "Methods for assessing the statistical significance of molecular sequence features by using general scoring schemes," *Proc. Natl. Acad. Sci. USA,* vol. 87, pp. 2264-2268, 1990.

[58] S.F. Altschul, "Amino acid substitution matrices from an information theoretic perspective," *J. Mol. Biol,* vol. 219, pp. 555-565, 1991.

[59] Z. Zhang, Schäffer, A.A., Miller, W., Madden, T.L., Lipman, D.J., Koonin, E.V., Altschul, S.F., "Protein sequence similarity searches using patterns as seeds," *Nucleic Acids Res,* vol. 26, pp. 3986-3990, 1998.

[60] O. Emanuelsson, Nielsen, H., Brunak, S., von Heijne, G., "Predicting subcellular localization of proteins based on their N-terminal amino acid sequence," *J Mol Biol,* vol. 300, pp. 1005-1016, 2000.

[61] J.L. Gardy, Spencer, C., Wang, K., Ester, M., Tusnády, G.E., Simon, I., Hua, S., deFays, K., Lambert, C., Nakai, K., Brinkman, F.S., "PSORT-B: Improving

protein subcellular localization prediction for Gram-negative bacteria," *Nucleic Acids Res,* vol. 31, pp. 3613-3617, 2003.

[62] D. Szafron, Lu, P., Greiner, R., Wishart, D.S., Poulin, B., Eisner, R., Lu, Z., Anvik, J., Macdonell, C., Fyshe, A., Meeuwis, D., "Proteome Analyst: custom predictions with explanations in a web-based tool for high-throughput proteome annotations," *Nucleic Acids Res,* vol. 32, pp. W365-71, 2004.

[63] F. Odronitz, Pillmann, H., Keller, O., Waack, S., Kollmar, M., "WebScipio: An online tool for the determination of gene structures using protein sequences," *BMC Genomics,* vol. 9, p. 422, 2008.

[64] C.R. Bradshaw, Surendranath, V., Habermann, B., "ProFAT: a web-based tool for the functional annotation of protein sequences," *BMC Bioinformatics,* vol. 7, p. 466, 2006.

[65] A.K. Nussbaum, Kuttler, C., Hadeler, K.P., Rammensee, H.G., Schild, H., "PAProC: a prediction algorithm for proteasomal cleavages available on the WWW," *Immunogenetics,* vol. 53, pp. 87-94, 2001.

[66] H. Singh, Raghava, G.P., "ProPred1: prediction of promiscuous MHC Class-I binding sites," *Bioinformatics,* vol. 19, pp. 1009-14, 2003.

[67] J. Tong, Jiang, P., Lu, Z.H., "RISP: a web-based server for prediction of RNA-binding sites in proteins," *Comput Methods Programs Biomed,* vol. 90, pp. 148-53, 2008.

[68] Y. Wang, Xue, Z., Shen, G., Xu, J., "PRINTR: prediction of RNA binding sites in proteins using SVM and profiles," *Amino Acids,* vol. 35, pp. 295-302, 2008.

[69] P.D. Taylor, Toseland, C.P., Attwood, T.K., Flower, D.R., "LIPPRED: A web server for accurate prediction of lipoprotein signal sequences and cleavage sites," *Bioinformation,* vol. 1, pp. 176-9, 2006.

[70] G. Su, Mao, B., Wang, J., "A web server for transcription factor binding site prediction," *Bioinformation,* vol. 1, pp. 156-7, 2006.

[71] K. Goyal, Mohanty, D., Mande, S.C., "PAR-3D: a server to predict protein active site residues," *Nucleic Acids Res,* vol. 35, pp. W503-5, 2007.

[72] D.T. Chang, Oyang, Y.J., Lin, J.H., "MEDock: a web server for efficient prediction of ligand binding sites based on a novel optimization algorithm," *Nucleic Acids Res,* vol. 33, pp. W233-8, 2005.

[73] Y. Zhang, "miRU: an automated plant miRNA target prediction server," *Nucleic Acids Res,* vol. 33, pp. W701-4, 2005.

[74] P. Duckert, Brunak, S. and Blom, N., "Prediction of proprotein convertase cleavage sites," *Protein Eng. Des. Sel,* vol. 17, pp. 107-113, 2004.

[75] E. Ferraro, Peluso, D., Via, A., Ausiello, G., Helmer-Citterich, M., "SH3-Hunter: discovery of SH3 domain interaction sites in proteins," *Nucleic Acids Res,* vol. 35, pp. W451-4, 2007.

[76] H.B. Shen, Chou, K.C., "Signal-3L: A 3-layer approach for predicting signal peptides," *Biochem Biophys Res Commun,* vol. 363, pp. 297-303, 2007.

[77] J.E. Hansen, Lund, O., Tolstrup, N., Gooley, A.A., Williams, K.L., Brunak, S., "NetOglyc: prediction of mucin type O-glycosylation sites based on sequence context and surface accessibility," *Glycoconj J*, vol. 15, pp. 115-120, 1998.

[78] C. Caragea, Sinapov, J., Silvescu, A., Dobbs, D., Honavar, V., "Glycosylation site prediction using ensembles of Support Vector Machine classifiers," *BMC Bioinformatics*, vol. 8, p. 438, 2007.

[79] N. Blom, Gammeltoft, S. and Brunak, S., "Sequence- and structure-based prediction of eukaryotic protein phosphorylation sites," *J Mol Biol*, vol. 294, pp. 1351-1362, 1999.

[80] F. Diella, Cameron, S., Gemund, C., Linding, R., Via, A., Kuster, B., Sicheritz-Ponten, T., Blom, N., Gibson, T.J., "Phospho.ELM: a database of experimentally verified phosphorylation sites in eukaryotic proteins," *BMC Bioinformatics*, vol. 22, p. 79, 2004.

[81] M. Hjerrild, Stensballe, A., Rasmussen, T.E., Kofoed, C.B., Blom, N., Sicheritz-Ponten, T., Larsen, M.R., Brunak, S., Jensen, O.N., Gammeltoft, S., "Identification of phosphorylation sites in protein kinase A substrates using artificial neural networks and mass spectrometry," *J Proteome Res*, vol. 3, pp. 426-433, 2004.

[82] C.R. Ingrell, Miller, M.L., Jensen, O.N., Blom, N., "NetPhosYeast: prediction of protein phosphorylation sites in yeast," *Bioinformatics*, vol. 23, pp. 895-7, 2007.

[83] L. Kiemer, Bendtsen, J.D. and Blom, N., "NetAcet: prediction of N-terminal acetylation sites," *Bioinformatics*, vol. 21, pp. 1269-1270, 2005.

[84] H. Chen, Xue, Y., Huang, N., Yao, X. and Sun, Z., "MeMo: a web tool for prediction of protein methylation modifications," *Nucleic Acids Res*, vol. 34, pp. W249-W253, 2006.

[85] Y. Xue, Zhou, F., Fu, C., Xu, Y. and Yao, X., "SUMOsp: a web server for sumoylation site prediction," *Nucleic Acids Res*, vol. 34, pp. W254-W257, 2006.

[86] Y. Xue, Chen, H., Jin, C., Sun, Z. and Yao, X., "NBA-Palm: prediction of palmitoylation site implemented in Naïve Bayes algorithm," *BMC Bioinformatics*, vol. 7, pp. 458-462, 2006.

[87] B. Eisenhaber, Bork, P., Eisenhaber, F., "Prediction of potential GPI-modification sites in proprotein sequences," *J Mol Biol*, vol. 292, pp. 741-58, 1999.

[88] Y. Wang, Xue, Z., Xu, J., "Better prediction of the location of alpha-turns in proteins with support vector machine," *Proteins*, vol. 65, pp. 49-54, 2006.

[89] A. Kirschner, Frishman, D., "Prediction of beta-turns and beta-turn types by a novel bidirectional Elman-type recurrent neural network with multiple output layers (MOLEBRNN)," *Gene*, vol. 422, pp. 22-9, 2008.

[90] Y. Wang, Xue, Z.D., Shi, X.H., Xu, J., "Prediction of pi-turns in proteins using PSI-BLAST profiles and secondary structure information," *Biochem Biophys Res Commun*, vol. 347, pp. 574-80, 2006.

[91] S. Costantini, Colonna, G., Facchiano, A.M., "PreSSAPro: a software for the prediction of secondary structure by amino acid properties," *Comput Biol Chem,* vol. 31, pp. 389-92, 2007.

[92] M. Duan, Huang, M., Ma, C., Li, L., Zhou, Y., "Position-specific residue preference features around the ends of helices and strands and a novel strategy for the prediction of secondary structures," *Protein Sci,* vol. 17, pp. 1505-12.

[93] R. Bondugula, Xu, D., "MUPRED: a tool for bridging the gap between template based methods and sequence profile based methods for protein secondary structure prediction," *Proteins,* vol. 66, pp. 664-70.

[94] S. Montgomerie, Sundararaj, S., Gallin, W.J., Wishart, D.S., "Improving the accuracy of protein secondary structure prediction using structural alignment," *BMC Bioinformatics,* vol. 7, p. 301, 2006.

[95] T.Z. Sen, Jernigan, R.L., Garnier, J., Kloczkowski, A., "GOR V server for protein secondary structure prediction," *Bioinformatics,* vol. 21, pp. 2787-8, 2005.

[96] G. Pollastri, McLysaght, A., "Porter: a new, accurate server for protein secondary structure prediction," *Bioinformatics,* vol. 21, pp. 1719-20, 2005.

[97] A.A. Salamov, Solovyev, V.V., "Protein secondary structure prediction using local alignments," *J Mol Biol,* vol. 268, pp. 31-6, 1997.

[98] P.Lieutaud, Canard, B., Longhi, S., "MeDor: a metaserver for predicting protein disorder," *BMC Genomics,* vol. 9, p. S25.

[99] D. Sethi, Garg, A., Raghava, G.P., "DPROT: prediction of disordered proteins using evolutionary information," *Amino Acids,* vol. 35, pp. 599-605, 2008.

[100] C.T. Su, Chen, C.Y., Hsu, C.M., "iPDA: integrated protein disorder analyzer," *Nucleic Acids Res,* vol. 35, pp. W465-72, 2007.

[101] T. Ishida, Kinoshita, K., "PrDOS: prediction of disordered protein regions from amino acid sequence," *Nucleic Acids Res,* vol. 35, pp. W460-4, 2007.

[102] O.V. Galzitskaya, Garbuzynskiy, S.O., Lobanov, M.Y., "FoldUnfold: web server for the prediction of disordered regions in protein chain," *Bioinformatics,* vol. 22, pp. 2948-0, 2006.

[103] A. Vullo, Bortolami, O., Pollastri, G., Tosatto, S.C., "Spritz: a server for the prediction of intrinsically disordered regions in protein sequences using kernel machines," *Nucleic Acids Res,* vol. 34, pp. W164-8, 2006.

[104] Z. Dosztányi, Csizmok, V., Tompa, P., Simon, I., "IUPred: web server for the prediction of intrinsically unstructured regions of proteins based on estimated energy content," *Bioinformatics,* vol. 21, pp. 3433-4, 2005.

[105] Z.R. Yang, Thomson, R., McNeil, P., Esnouf, R.M., "RONN: the bio-basis function neural network technique applied to the detection of natively disordered regions in proteins," *Bioinformatics,* vol. 21, pp. 3369-76, 2005.

[106] R. Linding, Jensen, L.J., Diella, F., Bork, P., Gibson, T.J., Russell, R.B., "Protein disorder prediction: implications for structural proteomics," *Structure,* vol. 11, pp. 1453-9, 2003.

[107] A. Campen, Williams, R.M., Brown, C.J., Meng, J., Uversky, V.N., Dunker, A.K., "TOP-IDP-scale: a new amino acid scale measuring propensity for intrinsic disorder," *Protein Pept Lett.*, vol. 15, pp. 956-63, 2008.

[108] R. Linding, Russell, R.B., Neduva, V., Gibson, T.J., "GlobPlot: Exploring protein sequences for globularity and disorder," *Nucleic Acids Res*, vol. 31, pp. 3701-8, 2003.

[109] P. Romero, Kissinger, C., Villafranca,J.E., Dunker, A.K., "Identifying disordered regions in proteins from amino acid sequence," in *Proc. IEEE Int. Conf. Neural Networks*, 1997, pp. 90-95.

[110] M. Khaladkar, Bellofatto, V., Wang, J.T., Tian, B., Shapiro, B.A., "RADAR: a web server for RNA data analysis and research," *Nucleic Acids Res*, vol. 35, pp. W300-4, 2007.

[111] W. Shu, Bo, X., Liu, R., Zhao, D., Zheng, Z., Wang, S., "RDMAS: a web server for RNA deleterious mutation analysis," *BMC Bioinformatics*, vol. 7, p. 404, 2006.

[112] J. Tárraga, Medina, I., Carbonell, J., Huerta-Cepas, J., Minguez, P., Alloza, E., Al-Shahrour, F., Vegas-Azcárate, S., Goetz, S., Escobar, P., Garcia-Garcia, F., Conesa, A., Montaner, D., Dopazo, J., "GEPAS, a web-based tool for microarray data analysis and interpretation," *Nucleic Acids Res*, vol. 36, pp. W308-14, 2008.

[113] G.E. Gonye, Chakravarthula, P., Schwaber, J.S., Vadigepalli, R., "From promoter analysis to transcriptional regulatory network prediction using PAINT," *Methods Mol Biol*, vol. 408, pp. 49-68, 2007.

[114] R. Diaz-Uriarte, "GeneSrF and varSelRF: a web-based tool and R package for gene selection and classification using random forest," *BMC Bioinformatics*, vol. 8, p. 328, 2007.

[115] I. Medina, Montaner, D., Tárraga, J., Dopazo, J., "Prophet, a web-based tool for class prediction using microarray data," *Bioinformatics*, vol. 23, pp. 390-1, 2007.

[116] M.A. Ott, Vriend, G., "Correcting ligands, metabolites, and pathways," *BMC Bioinformatics*, vol. 7, p. 517, 2006.

[117] T. Tokimatsu, Sakurai, N., Suzuki, H., Ohta, H., Nishitani, K., Koyama, T., Umezawa, T., Misawa, N., Saito, K., Shibata, D., "KaPPA-view: a web-based analysis tool for integration of transcript and metabolite data on plant metabolic pathway maps," *Plant Physiol*, vol. 138, pp. 1289-300, 2005.

[118] H. Goldstine, *The Computer: from Pascal to von Neumann*. Princeton, New Jersey: Princeton University Press, 1972.

[119] A. Michael, B., On the Way to the Web: The Secret History of the Internet and Its Founders. USA: Apress, 2008.

[120] T. Hastie, Tibshirani, R., Friedman, J., *The Elements of Statistical Learning*. New York: Springer, 2001.

[121] D.T. Chang, Wang, C.C., Chen, J.W., "Using a kernel density estimation based classifier to predict species-specific microRNA precursors," *BMC Bioinformatics*, vol. 12 p. S2, 2008.

[122] D.T. Chang, Ou, Y.Y., Hung, H.G., Yang, M.H., Chen, C.Y., Oyang, Y.J., "Prediction of protein secondary structures with a novel kernel density estimation based classifier," *BMC Res Notes*, vol. 1, p. 51, 2008.

[123] J. Gobeill, Tbahriti, I., Ehrler, F., Mottaz, A., Veuthey, A.L., Ruch, P., "Gene Ontology density estimation and discourse analysis for automatic GeneRiF extraction," *BMC Bioinformatics*, vol. 9, p. S9.

[124] T.B. Chen, Lu, H.H., Lee, Y.S., Lan, H.J., "Segmentation of cDNA microarray images by kernel density estimation," *J Biomed Inform*, vol. 41, pp. 1021-7, 2008.

[125] I.M. Overton, Padovani, G., Girolami, M.A., Barton, G.J., "ParCrys: a Parzen window density estimation approach to protein crystallization propensity prediction," *Bioinformatics*, vol. 24, pp. 901-7, 2008.

[126] J.D. Fischer, Mayer, C.E., Sading, J., "Prediction of protein functional residues from sequence by probability density estimation," *Bioinformatics*, vol. 24, pp. 613-20, 2008.

[127] K. Pearson, "On Lines and Planes of Closest Fit to Systems of Points in Space," *Philosophical Magazine*, vol. 2, pp. 559-572, 1901.

[128] F. Keinosuke, Introduction to Statistical Pattern Recognition: Elsevier, 1990.

[129] M. Cocchi, Durante, C., Grandi, M., Manzini, D., Marchetti, A., "Three-way principal component analysis of the volatile fraction by HS-SPME/GC of aceto balsamico tradizionale of modena," *Talanta*, vol. 74, pp. 547-54, 2008.

[130] W. Wu, Guo, Q., de Aguiar, P.F., Massart, D.L., "The star plot: an alternative display method for multivariate data in the analysis of food and drugs," *J Pharm Biomed Anal*, vol. 17, pp. 1001-13, 1998.

[131] J. Zhao, Patwa, T.H., Qiu, W., Shedden, K., Hinderer, R., Misek, D.E., Anderson, M.A., Simeone, D.M., Lubman, D.M., "Glycoprotein microarrays with multi-lectin detection: unique lectin binding patterns as a tool for classifying normal, chronic pancreatitis and pancreatic cancer sera.," *I Proteome Res*, vol. 6, pp. 1864-74, 2007.

[132] J. Albanese, Martens, K., Karanitsa, L.V., Schreyer, S.K., Dainiak, N., "Multivariate analysis of low-dose radiation-associated changes in cytokine gene expression profiles using microarray technology," *Exp Hematol*, vol. 35, pp. 47-54, 2007.

[133] Y.H. Taguchi, Oono, Y., "Relational patterns of gene expression via non-metric multidimensional scaling analysis," *Bioinformatics*, vol. 21, pp. 730-40, 2005.

[134] S.S. Yau, Yu, C., He, R., "A protein map and its application," *DNA Cell Biol*, vol. 27, pp. 241-50, 2008.

[135] A. Biegert, Mayer, C., Remmert, M., Söding, J., Lupas, A.N., "The MPI Bioinformatics Toolkit for protein sequence analysis," *Nucleic Acids Res*, vol. 34, pp. W335-9, 2006.

[136] S. Mayewski, "A multibody, whole-residue potential for protein structures, with testing by Monte Carlo simulated annealing," *Proteins*, vol. 59, pp. 152-69, 2005.

[137]   P. Li, Lu, X., Shao, M., Long, J., Wang, J., "Genetic diversity of harpins from Xanthomonas oryzae and their activity to induce hypersensitive response and disease resistance in tobacco," *Sci China C Life Sci,* vol. 47, pp. 461-9, 2004.

[138]   M. M. Bamman, Petrella, J.K., Kim, J.S., Mayhew, D.L., Cross, J.M., "Cluster analysis tests the importance of myogenic gene expression during myofiber hypertrophy in humans," *J Appl Physiol,* vol. 102, pp. 2232-9, 2007.

[139]   Y. Yamada, Fujimoto, A., Ito, A., Yoshimi, R., Ueda, M., "Cluster analysis and gene expression profiles: a cDNA microarray system-based comparison between human dental pulp stem cells (hDPSCs) and human mesenchymal stem cells (hMSCs) for tissue engineering cell therapy," *Biomaterials,* vol. 27, pp. 3766-81, 2006.

[140]   A. Tichopád, Pecen, L., Pfaffl, M.W., "Distribution-insensitive cluster analysis in SAS on real-time PCR gene expression data of steadily expressed genes," *Comput Methods Programs Biomed,* vol. 82, pp. 44-50, 2006.

[141]   A.E. Teschendorff, Wang, Y., Barbosa-Morais, N.L., Brenton, J.D., Caldas, C., "A variational Bayesian mixture modelling framework for cluster analysis of gene-expression data," *Bioinformatics,* vol. 21, pp. 3025-33, 2005.

[142]   X. Wu, Dewey, T.G., "Cluster analysis of dynamic parameters of gene expression," *J Bioinform Comput Biol,* vol. 1, pp. 447-58, 2003.

[143]   D.R. Bickel, "Robust cluster analysis of microarray gene expression data with the number of clusters determined biologically," *Bioinformatics,* vol. 19, pp. 818-24, 2003.

[144]   M.F. Ramoni, Sebastiani, P., Kohane, I.S., "Cluster analysis of gene expression dynamics," *Proc Natl Acad Sci U S A.,* vol. 99, pp. 9121-6, 2002.

[145]   T. Kohonen, *Self-Organizing Maps.* Berlin: Springer, 2001.

[146]   S.Y. Ku, Hu, Y.J., "Protein structure search and local structure characterization," *BMC Bioinformatics,* vol. 9, p. 349, 2008.

[147]   M. Meissner, Koch, O., Klebe, G., Schneider, G., "Prediction of turn types in protein structure by machine-learning classifiers," *Proteins,* vol. 74, pp. 344-52, 2009.

[148]   C. Martin, Diaz, N.N., Ontrup, J., Nattkemper, T.W., "Hyperbolic SOM-based clustering of DNA fragment features for taxonomic visualization and classification," *Bioinformatics,* vol. 24, pp. 1568-74, 2008.

[149]   J. Kim, Kim, J.H., "Difference-based clustering of short time-course microarray data with replicates," *BMC Bioinformatics,* vol. 8, p. 253, 2007.

[150]   K. Ning, Ng, H.K., Leong, H.W., "PepSOM: an algorithm for peptide identification by tandem mass spectrometry based on SOM," *Genome Inform,* vol. 17, pp. 194-205, 2006.

[151]   T. Ohlson, Aggarwal, V., Elofsson, A., MacCallum, R.M., "Improved alignment quality by combining evolutionary information, predicted secondary structure and self-organizing maps," *BMC Bioinformatics,* vol. 7, p. 357, 2006.

[152] J.M. Otaki, Mori, A., Itoh, Y., Nakayama, T., Yamamoto, H., "Alignment-free classification of G-protein-coupled receptors using self-organizing maps," *J Chem Inf Model*, vol. 46, pp. 1479-90, 2006.

[153] T. Abe, Sugawara, H., Kanaya, S., Kinouchi, M., Ikemura, T., "Self-Organizing Map (SOM) unveils and visualizes hidden sequence characteristics of a wide range of eukaryote genomes," *Gene*, vol. 365, pp. 27-34, 2006.

[154] X.S. Zhang, Wang, Y., Zhan, Z.W., Wu, L.Y., Chen, L., "Exploring protein's optimal HP configurations by self-organizing mapping," *J Bioinform Comput Biol*, vol. 3, pp. 385-400, 2005.

[155] M. Kanehisa, Goto, S., "KEGG: Kyoto Encyclopedia of Genes and Genomes," *Nucleic Acids Res*, vol. 28, pp. 27-30, 2000.

[156] E. Parzen, "On estimation of a probability density function and mode," *Ann. Math. Stat*, vol. 33, pp. 1065-76, 1962.

[157] A.P.L. Dempster, N.M.; Rubin, D.B., "Maximum Likelihood from Incomplete Data via the EM Algorithm," *Journal of the Royal Statistical Society. Series B (Methodological)*, vol. 39, pp. 1-38, 1977.

[158] R.O. Duda, Hart, P.E., Stork, D.G., *Pattern Classification*, 2nd ed.: Wiley-Interscience, 2000.

[159] C.M. Bishop, *Neural Networks for Pattern Recognition*. Bath, UK: Oxford University Press, USA, 1995.

[160] G.J. McLachlan, Discriminant Analysis and Statistical Pattern Recognition: Wiley Interscience, 2004.

[161] B.D. Ripley, *Pattern Recognition and Neural Networks*. Cambridge: Cambridge University Press, 1996.

[162] Y. Sakamoto, Ishiguro, M., and Kitagawa G., *Akaike Information Criterion Statistics*: D. Reidel Publishing Company, 1986.

[163] J. Kyte, Doolittle, R.F., "A simple method for displaying the hydropathic character of a protien," *J Mol Biol*, vol. 157, pp. 105-108, 1982.

[164] D.T. Chang, Wang, C.C., Chen, J.W., "Using a kernel density estimation based classifier to predict species-specific microRNA precursors," *BMC Bioinformatics*, vol. 12, p. S2, 2008.

[165] P. Mahata, Mahata, K., "Selecting differentially expressed genes using minimum probability of classification error," *J Biomed Inform*, vol. 40, pp. 775-86, 2007.

[166] I. Wasito, Hashim, S.Z., Sukmaningrum, S., "Iterative local Gaussian clustering for expressed genes identification linked to malignancy of human colorectal carcinoma," *Bioinformation*, vol. 2, pp. 175-81, 2007.

[167] X. Yan, Deng, M., Fung, W.K., Qian, M., "Detecting differentially expressed genes by relative entropy," *J Theor Biol*, vol. 234, pp. 395-402, 2005.

[168] I.T. Jolliffe, *Principal Component Analysis*. NY: Springer, 2002.

[169] D.G. Lemay, Neville, M.C., Rudolph, M.C., Pollard, K.S., German, J.B., "Gene regulatory networks in lactation: identification of global principles using bioinformatics," *BMC Syst Biol*, vol. 1, p. 56, 2007.

[170] J. Zhao, Patwa, T.H., Qiu, W., Shedden, K., Hinderer, R., Misek, D.E., Anderson, M.A., Simeone, D.M., Lubman, D.M., "Glycoprotein microarrays with multi-lectin detection: unique lectin binding patterns as a tool for classifying normal, chronic pancreatitis and pancreatic cancer sera.," *J Proteome Res,* vol. 6, pp. 1864-74, 2007.

[171] M. Szyma. A.E., M.J., Capron, X., van Nederkassel, A.M., Heyden, Y.V., Markuszewski, M., Krajka, K., Kaliszan, R., "Increasing conclusiveness of metabonomic studies by chem-informatic preprocessing of capillary electrophoretic data on urinary nucleoside profiles," *J Pharm Biomed Anal,* vol. 43, pp. 413-20, 2007.

[172] H. Fang, Xie, Q., Boneva, R., Fostel, J., Perkins, R., Tong, W., "Gene expression profile exploration of a large dataset on chronic fatigue syndrome," *Pharmacogenomics J,* vol. 7, pp. 429-40, 2006.

[173] J. Gao, Friedrichs, M.S., Dongre, A.R., Opiteck, G.J., "Guidelines for the routine application of the peptide hits technique," *J Am Soc Mass Spectrom,* vol. 16, pp. 1231-8, 2005.

[174] I.A. Doytchinova, Guan, P., Flower, D.R., "Identifiying human MHC supertypes using bioinformatic methods," *J Immunol,* vol. 172, pp. 4314-23, 2004.

[175] J.L. Griffin, Muller, D., Woograsingh, R., Jowatt, V., Hindmarsh, A., Nicholson, J.K., Martin, J.E., "Vitamin E deficiency and metabolic deficits in neuronal ceroid lipofuscinosis described by bioinformatics," *Physiol Genomics,* vol. 11, pp. 195-203, 2002.

[176] S.V. Edwards, Fertil, B., Giron, A., Deschavanne, P.J., "A genomic schism in birds revealed by phylogenetic analysis of DNA strings," *Syst Biol,* vol. 51, pp. 599-613, 2002.

[177] Z.R. Yang, Lertmemongkolchai, G., Tan, G., Felgner, P.L., Titball, R., "A Genetic Programming Approach for Burkholderia Pseudomallei Diagnostic Pattern Discovery," *Bioinformatics,* vol. in press, 2009.

[178] J.W. Sammon Jr, "A nonlinear mapping for data structure analysis," *IEEE Transactions on Computers,* vol. 18, pp. 401-9, 1969.

[179] P. Törönena, Kolehmainenb, M., Wonga, G., Castrén, E., "Analysis of gene expression data using self-organizing maps," *FEBS Lett,* vol. 451, pp. 142-6, 1999.

[180] R.M. Ewing, Cherry, J.M., "Visualization of expression clusters using Sammon's non-linear mapping," *Bioinformatics,* vol. 17, pp. 658-9, 2001.

[181] E. Niméus-Malmström, Krogh, M., Malmström,P., Strand, C., Fredriksson, I., Karlsson, P., Nordenskjöld, B., Stål, O., Östberg, G., Peterson, C., Fernö, M., "Gene expression profiling in primary breast cancer distinguishes patients developing local recurrence after breast-conservation surgery, with or without postoperative radiotherapy," *Breast Cancer Res,* vol. 10, p. R34, 2008.

[182] F. Azuaje, Wang, H., Chesneau, A., "Non-linear mapping for exploratory data analysis in functional genomics," *BMC Bioinformatics,* vol. 6, p. 13, 2005.

[183] T. Chen, Martin, E., Montague, G., "Robust probabilistic PCA with missing data and contribution analysis for outlier detection," *Computational Stat. & Data analysis,* vol. 53, pp. 3706-16, 2009.

[184] M.E. Tipping, Bishop, C.M., "Probabilistic principal component analysis," *Journal of the Royal Statistical Society. Series B (Statistical Methodology),* vol. 61, pp. 611-22, 2002.

[185] M. Scholz, Kaplan, F., Guy, C.L., Kopka, K., Selbig, J., "Non-linear PCA: a missing data approach," *Bioinformatics,* vol. 21, pp. 3887-95, 2005.

[186] J.G. Lees, Miles, A.J., Wien, F., Wallace, B.A., "A reference database for circular dichroism spectroscopy covering fold and secondary structure space," *Bioinformatics,* vol. 22, pp. 1955-62, 2006.

[187] F.M. Selaru, Zou, T., Xu, Y., Shustova, V., Yin, J., Mori, Y., Sato, F., Wang, S., Olaru, A., and D. Shibata, Greenwald, B.D., Krasna, M.J., Abraham, J.M., Meltzer, S.J., "Global gene expression profiling in Barrett's esophagus and esophageal cancer: a comparative analysis using cDNA microarrays," *Oncogene,* vol. 21, pp. 475-8, 2002.

[188] J.C. Dunn, "A Fuzzy Relative of the ISODATA Process and Its Use in Detecting Compact Well-Separated Clusters," *Journal of Cybernetics,* vol. 3, pp. 32-57, 1973.

[189] J.C. Bezdek, Pattern Recognition with Fuzzy Objective Function Algorithms. NY: Plenum, 1981.

[190] T.D.A. Smith, Makov, U., *Statistical Analysis of Finite Mixture Distributions*: John Wiley & Sons, 1985.

[191] B.S. Everitt, Hand D.J., *Finite mixture distributions*: Chapman & Hall, 1981.

[192] G.J. McLachlan, Peel, D., *Finite Mixture Models*: Wiley 2000.

[193] A. Gersho, "Asymptotically optimal block quantization," *IEEE Trans. Inform. Theory,* vol. IT-25, pp. 373-80, 1979.

[194] Y. Linde, Buzo, A., Gray, R.M., "An algorithm for vector quantizer design," *IEEE Transactions on Communications,* vol. 28, pp. 84-95, 1980.

[195] J. Makhoul, Roucos, S., Gish, H., "Vector Quantization in Speech Coding," *Proc. IEEE,* vol. 73, pp. 1551-88, 1985.

[196] P. Zador, "Asymptotic Quantization Error of Continuous Signals and the Quantization Dimension," *IEEE Trans. Inform. Theory,* vol. IT-28, pp. 139-49, 1982.

[197] C. von der Malsburg, "Self-organization of orientation sensitive cells in the striate cortex," *Kybernetik,* vol. 14, pp. 85-100, 1973.

[198] T. Kohonen, "Analysis of a simple self-organizing process," *Biological Cybernetics,* vol. 44, pp. 135-40, 1982.

[199] R.S. Istepanian, Sungoor, A., Nebel, J.C., "Fractal dimension and wavelet decomposition for robust microarray data clustering," *Conf Proc IEEE Eng Med Biol Soc,* pp. 4106-9, 2008.

[200] J. Li, Zha, H., "Simultaneous classification and feature clustering using discriminant vector quantization with applications to microarray data analysis," *Proc IEEE Comput Soc Bioinform Conf,* vol. 1, pp. 246-55, 2002.

[201] J. Hanke, Beckmann, G., Bork, P., Reich, J.G., "Self-organizing hierarchic networks for pattern recognition in protein sequence," *Protein Sci,* vol. 5, pp. 72-82, 1996.

[202] T. Abe, Kanaya, S., Kinouchi, M., Ichiba, Y., Kozuki, T., Ikemura, T., "Informatics for unveiling hidden genome signatures," *Genome Research,* vol. 13, pp. 693-702, 2003.

[203] P. Törönen, Kolehmainen, M., Wong, G., Castrén, E., "Analysis of gene expression data using self-organizing maps," *FEBS Lett,* vol. 451, pp. 142-6, 1999.

[204] S. Kanaya, Kinouchi, M., Abe, T., Kudo, Y., Yamada, Y., Nishi, T., Mori, H., Ikemura, T., "Analysis of codon usage diversity of bacterial genes with a self-organizing map (SOM): characterization of horizontally transferred genes with emphasis on the E. coli O157 genome," *Gene,* vol. 276, pp. 89-99, 2001.

[205] C. Martin, Diaz, N.N., Ontrup, J., Nattkemper, T.W., "Hyperbolic SOM-based clustering of DNA fragment features for taxonomic visualization and classification," *Bioinformatics,* vol. 24, pp. 1568-74, 2008.

[206] J. Wang, Delabie, J., Aasheim, H., Smeland, E., Myklebost, O., "Clustering of the SOM easily reveals distinct gene expression patterns: results of a reanalysis of lymphoma study," *BMC Bioinformatics,* vol. 3, p. 36, 2002.

[207] T.K. Baker, Carfagna, M.A., Gao, H., Dow, E.R., Li, Q., Searfoss, G.H., Ryan, T.P., "Temporal gene expression analysis of monolayer cultured rat hepatocytes," *Chem Res Toxicol,* vol. 14, pp. 1218-31, 2001.

[208] Y.D. Cai, Yu, H., Chou, K.C., "Artificial neural network method for predicting HIV protease cleavage sites in protein," *J Protein Chem,* vol. 17, pp. 607-15, 1998.

[209] N. Qian, Sejnowski, T., "Predicting the secondary structure of globu-lar proteins using neural network models," *J Mol Biol. Aug 20;(4):,* vol. 202, pp. 865-84, 1988.

[210] S. Tümpel, Maconochie, M., Wiedemann, L.M., Krumlauf, R., "Conservation and Diversity in the cis-Regulatory Networks That Integrate Information Controlling Expression of Hoxa2 in Hindbrain and Cranial Neural Crest Cells in Vertebrates," *Developmental Biology,* vol. 246, pp. 45-56, 2002.

[211] P.F. Kemp, Aller, J.Y., "Bacterial diversity in aquatic and other environments: what 16S rDNA libraries can tell us," *FEMS Microbiology Ecology,* vol. 47, pp. 161-77, 2004.

[212] A.C. Lorena, de Carvalho, A.C.P.L.F., "Protein cellular localization prediction with Support Vector Machines and Decision Trees," *Computers in Biology and Medicine,* vol. 37, pp. 115-25, 2007.

[213] A. Ruepp, Mewes, H.W., "Prediction and classification of protein functions," *Drug Discovery Today: Technologies,* vol. 3, pp. 145-51, 2006.

[214] B.H. Dessailly, Redfern, O.C., Cuff, A., Orengo, C.A., "Exploiting structural classifications for function prediction: towards a domain grammar for protein function," *Current Opinion in Structural Biology,* vol. 19, pp. 349-56, 2009.

[215] Y. Gusev, "Computational methods for analysis of cellular functions and pathways collectively targeted by differentially expressed microRNA," *Methods,* vol. 44, pp. 61-72, 2008.

[216] O. Emanuelsson, Nielsen, H., von Heijne, G., "ChloroP, a neural network-based method for predicting chloroplast transit peptides and their cleavage sites," *Protein Science,* vol. 8, pp. 978-984, 1999.

[217] A.S. Juncker, Willenbrock, H., von Heijne, G., Nielsen, H., Brunak, S., Krogh, A., "Prediction of lipoprotein signal peptides in Gram-negative bacteria," *Protein Sci,* vol. 12, pp. 1652-62, 2003.

[218] M.G. Claros, Vincens, P., "Computational method to predict mitochondrially imported proteins and their targeting sequences," *Eur. J. Biochem,* vol. 241, pp. 779-786, 1996.

[219] J. Zuegge, Ralph, S., Schmuker, M., McFadden, G.I., Schneider, G., "Deciphering apicoplast targeting signals - feature extraction from nuclear-encoded precursors of Plasmodium falciparum apicoplast proteins," *Gene* vol. 280, pp. 19-26, 2001.

[220] A. Bender, van Dooren, G.G., Ralph, S.A., McFadden, G.I., Schneider, G., "Properties and prediction of mitochondrial transit peptides from Plasmodium falciparum," *Mol. Biochem. Parasitol,* vol. 132, pp. 59-66, 2003.

[221] I. Small, Peeters, N., Legeai, F., Lurin, C., "Predotar: A tool for rapidly screening proteomes for N-terminal targeting sequences," *Proteomics,* vol. 4, pp. 1581-1590, 2004.

[222] G. Neuberger, Maurer-Stroh, S., Eisenhaber, B., Hartig, A., Eisenhaber, F., "Motif refinement of the peroxisomal targeting signal 1 and evaluation of taxon-specific differences," *J Mol Biol,* vol. 328, pp. 567-79, 2003.

[223] J.D. Bendtsen, Nielsen, H., von Heijne, G., Brunak, S., "Improved prediction of signal peptides: SignalP 3.0.," *J. Mol. Biol,* vol. 340, pp. 783-795, 2004.

[224] R. Gupta, Jung, E., Gooley, A.A., Williams, K.L., Brunak, S., Hansen, J., "Scanning the available Dictyostelium discoideum proteome for O-linked GlcNAc glycosylation sites using neural networks," *Glycobiology,* vol. 9, pp. 1009-22, 1999.

[225] K. Julenius, "NetCGlyc 1.0: Prediction of mammalian C-mannosylation sites," *Glycobiology,* vol. 17, pp. 868-876, 2007.

[226] K. Julenius, Mølgaard, A., Gupta, R., Brunak, S., "Prediction, conservation analysis and structural characterization of mammalian mucin-type O-glycosylation sites," *Glycobiology,* vol. 15, pp. 153-164, 2005.

[227] M.B. Johansen, Kiemer, L., Brunak, S., "Analysis and prediction of mammalian protein glycation," *Glycobiology,* vol. 16, pp. 844-853, 2006.

[228] R. Gupta, Brunak, S., "Prediction of glycosylation across the human proteome and the correlation to protein function," *Pacific Symposium on Biocomputing,* vol. 7, pp. 310-322, 2002.

[229] B. Eisenhaber, Bork, P., Eisenhaber, F., "Sequence properties of GPI-anchored proteins near the omega-site: constraints for the polypeptide binding site of the putative transamidase," *Protein Engineering,* vol. 11, pp. 1155-1161, 1998.

[230] G. Bologna, Yvon, C., Duvaud, S., Veuthey, A.L., "N-terminal Myristoylation Predictions by Ensembles of Neural Networks," *Proteomics,* vol. in press, 2009.

[231] J. Ren, Wen, L., Gao, X., Jin, C., Xue, Y., Yao, X., "CSS-Palm 2.0: an updated software for palmitoylation sites prediction," *Protein Engineering, Design and Selection,* vol. 21, pp. 639-644, 2008.

[232] L. Kiemer, Bendtsen, J.D., Blom, N., "NetAcet: Prediction of N-terminal acetylation sites," *Bioinformatics,* vol. 21, pp. 1269-1270, 2005.

[233] N. Blom, Gammeltoft, S., and Brunak, S., "Sequence- and structure-based prediction of eukaryotic protein phosphorylation sites," *J Mol Biol,* vol. 294, pp. 1351-1362, 1999.

[234] N. Blom, Sicheritz-Ponten, T., Gupta, R., Gammeltoft, S., Brunak, S., "Prediction of post-translational glycosylation and phosphorylation of proteins from the amino acid sequence," *Proteomics,* vol. 4, pp. 1633-49.

[235] C.R. Ingrell, Miller, M.L., Jensen, O.N., Blom, N., "NetPhosYeast: Prediction of protein phosphorylation sites in yeast," *Bioinformatics,* vol. in press, 2009.

[236] N. Blom, Hansen,J., Blaas, D., Brunak, S., "Cleavage site analysis in picornaviral polyproteins: Discovering cellular targets by neural networks," *Protein Science,* vol. 5, pp. 2203-2216, 1996.

[237] L. Kiemer, Lund, O., Brunak, S., Blom, N., "Coronavirus 3CL-pro proteinase cleavage sites: Possible relevance to SARS virus pathology," *BMC Bioinformatics,* vol. 7, p. 72, 2004.

[238] P. Duckert, Brunak, S., Blom, S., "Prediction of proprotein convertase cleavage sites," *Protein Engineering, Design and Selection,* vol. 17, pp. 107-112, 2004.

[239] K. Krause, Eszlinger, M., Gimm, O., Karger, S., Engelhardt, C., Dralle, H., Fuhrer, D., "TFF3-based candidate gene discrimination of benign and malignant thyroid tumors in a region with borderline iodine deficiency," *J Clin Endocrinol Metab,* vol. 93, pp. 1390-3, 2008.

[240] J.A. Cruz, Wishart, D.S., "Applications of machine learning in cancer prediction and prognosis," *Cancer Inform,* vol. 2, pp. 59-77, 2007.

[241] T. Bellotti, Luo, Z., Gammerman, A., Van Delft, F.W., Saha, V., "Qualified predictions for microarray and proteomics pattern diagnostics with confidence machines," *Int J Neural Syst,* vol. 15, pp. 247-58, 2005.

[242] F. Li, Yang, Y., "Analysis of recursive gene selection approaches from microarray data," *Bioinformatics,* vol. 21, pp. 3714-7, 2005.

[243] E. Vairaktaris, Yapijakis, C., Serefoglou, Z., Avgoustidis, D., Critselis, E., Spyridonidou, S., Vylliotis, A., Derka, S., Vassiliou, S., Nkenke, E., Patsouris, E.,

"Gene expression polymorphisms of interleukins-1 beta, -4, -6, -8, -10, and tumor necrosis factors-alpha, -beta: regression analysis of their effect upon oral squamous cell carcinoma," *J Cancer Res Clin Oncol,* vol. 134, pp. 821-32, 2008.

[244] M. Steinfath, Groth, D., Lisec, J., Selbig, J., "Metabolite profile analysis: from raw data to regression and classification," *Physiol Plant,* vol. 132, pp. 150-61, 2008.

[245] H.G. Müller, Chiou, J.M., Leng, X., "Inferring gene expression dynamics via functional regression analysis," *BMC Bioinformatics,* vol. 9, p. 60, 2008.

[246] R.A. Barkley, Grodzinksi, G.M., "Are tests of frontal lobe functions useful in the diagnosis of Attention Deficit Disorders?," *The Clinical Neurologist,* vol. 8, pp. 121-39, 1994.

[247] R.W. Ellwood, "Clinical discriminations and neuropsychological tests: An appeal to Bayes' theorem," *The Clinical Neuropsychologist,* vol. 7, pp. 224-33, 1993.

[248] K. Matier-Sharma, Perachio, N., Newcorn, J.H., Sharma, V., Halperin, J.M., "Differential diagnosis of ADHD: Are objective measures of attention, impulsivity, and activity level helpful?," *Child Neuropsyehology,* vol. 1, pp. 118-127, 1995.

[249] J.N. Wherry, Paal, N., Jolly, J.B., Balkozar, A., Holloway, C., Everett, B., Vaught, L., "Concurrent and discriminant validity of the Gordon Diagnostic System: A preliminary study," *Psychology in the Schools,* vol. 1, pp. 29-36, 1993.

[250] C.E. Metz, "Basic principles of ROC analysis," *Seminars in Nuclear Medicine,* vol. 8, pp. 283-288, 1978.

[251] T. Sing, Sander, O., Beerenwinkel, N., Lengauer, T., "ROCR: visualizing classifier performance in R," *Bioinformatics,* vol. 21, pp. 3940-1, 2005.

[252] B. Efron, "Nonparametric estimates of standard error: The jackknife, the bootstrap and other methods," *Biometrika,* vol. 68, pp. 589-599, 1981.

[253] R.A. Fisher, "The use of multiple measurements in taxonomic problems," *Annals of Eugenics,* vol. 7, pp. 179-88, 1936.

[254] W. Dai, Teodoridis, J.M., Graham, J., Zeller, C., Huang, T.H., Yan, P., Vass, J.K., Brown, R., Paul, J., "Methylation Linear Discriminant Analysis (MLDA) for identifying differentially methylated CpG islands," *BMC Bioinformatics,* vol. 9, p. 337, 2008.

[255] X. Du, Yang, F., Manes, N.P., Stenoien, D.L., Monroe, M.E., Adkins, J.N., States, D.J., Purvine, S.O., Camp, D.G., Smith, R.D., "Linear discriminant analysis-based estimation of the false discovery rate for phosphopeptide identifications," *J Proteome Res,* vol. 7, pp. 2195-203, 2008.

[256] K.S. Opstad, Ladroue, C., Bell, B.A., Griffiths, J.R., Howe, F.A., "Linear discriminant analysis of brain tumour (1)H MR spectra: a comparison of classification using whole spectra versus metabolite quantification," *NMR Biomed,* vol. 20, pp. 763-70.

[257] J. Jin, "Identification of protein coding regions of rice genes using alternative spectral rotation measure and linear discriminant analysis," *Genomics Proteomics Bioinformatics,* vol. 2, pp. 167-73, 2004.

[258] M.Q. Zhang, "Identification of protein coding regions in the human genome by quadratic discriminant analysis," *PNAS,* vol. 94, pp. 565-8, 1997.

[259] L. Zhang, Luo, L., "Splice site prediction with quadratic discriminant analysis using diversity measure," *NAR,* vol. 31, pp. 6214-20, 2003.

[260] W. Chen, Luo, L., "Classification of antimicrobial peptide using diversity measure with quadratic discriminant analysis," *J Microbiol Methods,* vol. 78, pp. 94-6, 2009.

[261] A.G. Garrow, Agnew, A., Westhead, D.R., "TMB-Hunt: a web server to screen sequence sets for transmembrane beta-barrel proteins," *NAR,* vol. 33, pp. W188-92, 2005.

[262] A. Zorzet, Gustafsson, M., Hammerling, U., "Prediction of food protein allergenicity: a bioinformatic learning systems approach," *In Silico Biology,* vol. 2, pp. 525-34, 2002.

[263] A.C. Tan, Naiman, D.Q., Xu, L., Winslow, R.L., Geman, D., "Simple decision rules for classifying human cancers from gene expression profiles," *Bioinformatics,* vol. 21, pp. 3896-904, 2005.

[264] A. Barrier, Lemoine, A., Boelle, P.Y., Tse, C., Brault, D., Chiappini, F., Breittschneider, J., Lacaine, F., Houry, S., Huguier, M., Van der Laan, M.J., Speed, T., Debuire, B., Flahault, A, Dudoit, S., "Colon cancer prognosis prediction by gene expression profiling," *Oncogene,* vol. 24, pp. 6155-64, 2005.

[265] J. Li, Liu, H., Downing, J.R., Yeoh, A.E., Wong, L., "Simple rules underlying gene expression profiles of more than six subtypes of acute lymphoblastic leukemia (ALL) patients," *Bioinformatics,* vol. 19, pp. 71-8, 2003.

[266] J.H. van Delft, van Agen, E., van Breda, S.G., Herwijnen, M.H., Staal, Y.C., Kleinjans, J.C., "Comparison of supervised clustering methods to discriminate genotoxic from non-genotoxic carcinogens by gene expression profiling," *Mutat Res,* vol. 575, pp. 17-33, 2005.

[267] H. Wiesinger-Mayr, Vierlinger, K., Pichler, R., Kriegner, A., Hirschl, A.M., Presterl, E., Bodrossy, L., Noehammer, C., "Identification of human pathogens isolated from blood using microarray hybridisation and signal pattern recognition," *BMC Microbiol,* vol. 7, p. 78, 2007.

[268] B.C. Emerson, Kolm, N., "Species diversity can drive speciation," *Nature,* vol. 434, pp. 1015-1017, 2005.

[269] J.R. Quinlan, *C4.5: Programs for Machine Learning*: Morgan Kaufmann Publishers, 1993.

[270] J.R. Quinlan, "Improved use of continuous attributes in c4.5," *Journal of Artificial Intelligence Research,* vol. 4, pp. 77-100, 1996.

[271] L. Breiman, Friedman, J.H., Olshen, R.A., Stone, C.J., *Classification and Regression Trees*: Wadsworth, 1984.

[272] L. Breiman, "Random forests," *Machine Learning,* vol. 45, pp. 5-32, 2001.

[273] F. Hammann, Gutmann, H., Jecklin, U., Maunz, A., Helma, C., Drewe, J., "Development of decision tree models for substrates, inhibitors, and inducers of p-glycoprotein," *Curr Drug Metab,* vol. 10, pp. 339-46, 2009.

[274] S. Sethi, Benninger, M.S., Lu, M., Havard, S., Worsham, M.J., "Noninvasive molecular detection of head and neck squamous cell carcinoma: an exploratory analysis," *Diagn Mol Pathol,* vol. 18, pp. 81-7, 2009.

[275] M. García-Magariños, López-de-Ullibarri, I., Cao, R., Salas, A., "Evaluating the ability of tree-based methods and logistic regression for the detection of SNP-SNP interaction," *Ann Hum Genet,* vol. 73, pp. 360-9, 2009.

[276] H. Kulkarni, Agan, B.K., Marconi, V.C., O'Connell, R.J., Camargo, J.F., He, W., Delmar, J., Phelps, K.R., Crawford, G., Clark, R.A., Dolan, M.J., Ahuja, S.K., "CCL3L1-CCR5 genotype improves the assessment of AIDS Risk in HIV-1-infected individuals," *PLoS One,* vol. 3, p. e3165, 2008.

[277] H. Yang, Lippman, S.M., Huang, M., Lee. J., Wang, W., Spitz, M.R., Wu, X., "Genetic polymorphisms in double-strand break DNA repair genes associated with risk of oral premalignant lesions," *Eur J Cancer,* vol. 44, pp. 1603-11, 2008.

[278] Y. Ye, Yang, H., Grossman, H.B., Dinney, C., Wu, X., Gu, J., "Genetic variants in cell cycle control pathway confer susceptibility to bladder cancer," *Cancer,* vol. 112, pp. 2467-74, 2008.

[279] U. Raju, Mei, L., Seema, S., Hina, Q., Wolman, S.R., Worsham, M.J., "Molecular classification of breast carcinoma in situ," *Cuur Genomics,* vol. 7, pp. 523-32, 2006.

[280] A. Papana, Ishwaran, H., "CART variance stabilization and regularization for high-throughput genomic data," *Bioinformatics,* vol. 22, pp. 2254-61, 2006.

[281] R.D. Loss, "Atomic weights of the elements 2001," *Pure Appl. Chem.,* vol. 75, pp. 1107-22, 2003.

[282] B. Efron, Halloran‡, E., Holmes, S., "Bootstrap confidence levels for phylogenetic trees," *PNAS* vol. 93, p. 23, 1996.

[283] O. Tastan, Qi, Y., Carbonell, J.G., Klein-Seetharaman, J., "Prediction of interactions between HIV-1 and human proteins by information integration," *Pac Symp Biocomput,* pp. 516-27, 2009.

[284] X.Y. Wu, Wu, Z.Y., Li, K., "Identification of differential gene expression for microarray data using recursive random forest," *Chin Med J,* vol. 121, pp. 2492-6, 2008.

[285] J. Olsen, Gerds, T.A., Seidelin, J.B., Csillag, C., Bjerrum, J.T., Troelsen, J.T., Nielsen, O.H., "Diagnosis of ulcerative colitis before onset of inflammation by multivariate modeling of genome-wide gene expression data," *Inflamm Bowel Dis,* vol. in press, 2009.

[286] M.C. Abba, Sun, H., Hawkins, K.A., Drake, J.A., Hu, Y., Nunez, M.I., Gaddis, S., Shi, T., Horvath, S., Sahin, A., Aldaz, C.M., "Breast cancer molecular signatures

as determined by SAGE: correlation with lymph node status," *Mol Cancer Res,* vol. 5, pp. 881-90, 2007.

[287] K. Hoffmann, Firth, M.J., Beesley, A.H., de Klerk, N.H., Kees, U.R., "Translating microarray data for diagnostic testing in childhood leukaemia," *BMC Cancer,* vol. 6, p. 229, 2006.

[288] Y. Qi, Bar-Joseph, Z., Klein-Seetharaman, J., "Evaluation of different biological data and computational classification methods for use in protein interaction prediction," *Proteins,* vol. 63, pp. 490-500, 2006.

[289] W. McCulloch, Pitts, W., "A logical calculus the ideas immanent in nervous activity," *Bulletin Mathematical Biophysics,* vol. 5, pp. 115-33, 1943.

[290] F.B. Fitch, McCulloch, W.S., Pitts, W., "A logic calculus of the ideas immanent in nervous activity," *Journal Symbolic Logic,* vol. 9, pp. 49-50, 1944.

[291] D.O. Hebb, *The organization of behaviour*: John Wiley and Sons Inc, 1949.

[292] E. A. Feigenbaum, Feldman, J., "Computers & Thought," Cambridge, MA, USA: MIT Press, 1954.

[293] M. Minksy, *Perceptron*. Cambridge, MA, USA: MIT Press, 1969.

[294] J.J. Hopfield, "Neural networks and physical systems with emergent collective computational abilities," *PNSA,* vol. 79, pp. 2554-8, 1982.

[295] P.J. Werbos, The Roots of Backpropagation: From Ordered Derivatives to Neural Networks and Political Forecasting: Willey-Interscience, 1994.

[296] D.E. Rumelhart, McClelland, J.L, *Parallel Dsitributed Processing*: MIT press, 1986.

[297] D. Hecht, Cheung, M., Fogel, G.B., "QSAR using evolved neural networks for the inhibition of mutant PfDHFR by pyrimethamine derivatives," *Biosystems,* vol. 92, pp. 10-5, 2008.

[298] B. Slabbinck, De Baets, B., Dawyndt, P., De Vos, P., "Genus-wide Bacillus species identification through proper artificial neural network experiments on fatty acid profiles," *Antonie Van Leeuwenhoek,* vol. 94, pp. 187-98, 2008.

[299] M. Spreafico, Boriani, E., Benfenati, E., Novic, M., "Structural features of diverse ligands influencing binding affinities to estrogen alpha and estrogen beta receptors. Part II. Molecular descriptors calculated from conformation of the ligands in the complex resulting from previous docking study," *Mol Divers,* vol. 13, pp. 171-81, 2008.

[300] M. Jalali-Heravi, Asadollahi-Baboli, M., Shahbazikhah, P., "QSAR study of heparanase inhibitors activity using artificial neural networks and Levenberg-Marquardt algorithm," *Eur J Med Chem,* vol. 43, pp. 548-56, 2008.

[301] D. Catchpoole, Lail, A., Guo, D., Chen, Q.R., Khan, J., "Gene expression profiles that segregate patients with childhood acute lymphoblastic leukaemia: an independent validation study identifies that endoglin associates with patient outcome," *Leuk Res,* vol. 31, pp. 1741-7, 2007.

[302] C.L. Moore, Smagala, J.A., Smith, C.B., Dawson, E.D., Cox, N.J., Kuchta, R.D., Rowlen, K.L., "Evaluation of MChip with historic subtype H1N1 influenza A

viruses, including the 1918 "Spanish Flu" strain," *Journal Clinic Microbiology,* vol. 45, pp. 3807-10, 2007.

[303] R. Xu, Venayagamoorthy, G.K., Wunsch, D.C., "Modeling of gene regulatory networks with hybrid differential evolution and particle swarm optimization," *Neural Networks,* vol. 20, pp. 917-27, 2007.

[304] J.H. Chiang, Chao, S.Y., "Modeling human cancer-related regulatory modules by GA-RNN hybrid algorithms," *BMC Bioinformatics,* vol. 8, p. 91, 2007.

[305] S. Wagner, Arce, R., Murillo, R., Terfloth, L., Gasteiger, J., Merfort, I., "Neural networks as valuable tools to differentiate between sesquiterpene lactones' inhibitory activity on serotonin release and on NF-kappaB," *J Med Chem,* vol. 51, pp. 1324-32, 2008.

[306] M. Cruz-Monteagudo, Cordeiro, M.N., Borges, F., "Computational chemistry approach for the early detection of drug-induced idiosyncratic liver toxicity," *Journal Computer Chemistry,* vol. 29, pp. 533-49, 2008.

[307] L. Wang, Zheng, W., Mu, L., Zhang, S.Z., "Identifying biomarkers of endometriosis using serum protein fingerprinting and artificial neural networks," *International Journal Gynaecol Obstet,* vol. 101, pp. 253-8, 2008.

[308] Z. Zhang, Yu, Y., Xu, F., Berchuck, A., van Haaften-Day, C., Havrilesky, L.J., de Bruijn, H.W., van der Zee, A.G., Woolas, R.P., Jacobs, I.J., Skates, S., Chan, D.W., Bast, R.C., "Combining multiple serum tumor markers improves detection of stage I epithelial ovarian cancer," *Gynecol Oncol,* vol. 107, pp. 526-31., 2007.

[309] Y. Matsubara, Kikuchi, S., Sugimoto, M., Tomita, M., "Parameter estimation for stiff equations of biosystems using radial basis function networks," *BMC Bioinformatics,* vol. 7, p. 230, 2006.

[310] J.H. Chiang, Ho, S.H., "A combination of rough-based feature selection and RBF neural network for classification using gene expression data," *IEEE Trans Nanobioscience,* vol. 7, pp. 91-9, 2008.

[311] C.S. Möller-Levet, Yin, H., "Modeling and analysis of gene expression time-series based on co-expression," *Int J Neural Syst,* vol. 15, pp. 311-22, 2005.

[312] H.Q. Wang, Huang, D.S., "Non-linear cancer classification using a modified radial basis function classification algorithm," *J Biomed Sci,* vol. 12, pp. 819-26, 2005.

[313] L.E. Peterson, Coleman, M.A., "Machine learning-based receiver operating characteristic (ROC) curves for crisp and fuzzy classification of DNA microarrays in cancer research," *Int J Approx Reason,* vol. 47, pp. 17-36, 2008.

[314] N. Pochet, De Smet, F., Suykens, J.A., De Moor, B.L., "Systematic benchmarking of microarray data classification: assessing the role of non-linearity and dimensionality reduction," *Bioinformatics,* vol. 20, pp. 3185-95, 2004.

[315] S. Takasaki, Kawamura, Y., Konagaya, A., "Selecting effective siRNA sequences by using radial basis function network and decision tree learning," *BMC Bioinformatics,* vol. Suppl 5, p. S22, 2006.

[316] Y. Xue, Li, A., Wang, L., Feng, H., Yao, X., "PPSP: prediction of PK-specific phosphorylation site with Bayesian decision theory," *BMC Bioinformatics,* vol. 7, p. 163, 2006.

[317] Z.R. Yang, Thomson, R., "Bio-basis function neural network for prediction of protease cleavage sites in proteins," *IEEE Trans. on Neural Networks,* vol. 15, pp. 263-274, 2005.

[318] R. Thomson, Hodgman, T., Yang, Z.R. and Doyle, A., "Characterising proteolytic cleavage site activity using bio-basis function neural networks," *Bioinformatics,* vol. 19, pp. 1741-1447, 2003.

[319] E. Berry, Dalby, A. and Yang, Z.R., "Reduced bio basis function neural network for identification of protein phosphorylation sites: Comparison with pattern recognition algorithms," *Computational Biology and Chemistry* vol. 28, pp. 75-85, 2004.

[320] R. Thomson, Esnouf, R., "Predict disordered proteins using bio-basis function neural networks," *Lecture Notes in Computer Science,* vol. 3177, pp. 19-27, 2004.

[321] P. Senawongse, Dalby, A., Yang, Z.R., "Predicting the phosphorylation sites using hidden Markov models and Machine Learning methods," *Journal of Chemical Information and Computer Science,* vol. 45, pp. 1147-52, 2005.

[322] Z.R. Yang, Chou, K.C., "Predicting the linkage sites in glycoproteins using bio-basis function neural netwoek," *Bioinformatics,* vol. 20, pp. 903-8, 2004.

[323] A. Sidhu, Yang, Z.R., "Predict signal peptides using bio-basis function neural networks," *Applied Bioinformatics,* vol. 5, pp. 13-9, 2006.

[324] Z.R. Yang, Dry, J., Thomson, R. Hodgman, T.C., "A bio-basis function neural network for protein peptide cleavage activity characterisation," *Neural Networks,* vol. 19, pp. 401-407, 2006.

[325] Z.R. Yang, Young, N., "Bio-kernel Self-organizing map for HIV drug resistance classification," *Lecture Notes in Computer Science,* vol. 3610, pp. 179-84, 2005.

[326] Z.R. Yang, "Prediction of caspase cleavage sites using Bayesian bio-basis function neural networks," *Bioinformatics,* vol. 21, pp. 1831-7, 2005

[327] Z.R. Yang, "Mining SARS-CoV protease cleavage data using decision trees, a novel method for decisive template searching," *Bioinformatics,* vol. 21, pp. 2644-50, 2005.

[328] Z.R. Yang, Johnathan, F., "Predict T-cell epitopes using bio-support vector machines," *Journal of Chemical Informatics and Computer Science,* vol. 45, pp. 1142-8, 2005.

[329] K.A. Sepkowitz, "AIDS--the first 20 years," *N Engl J Med,* vol. 344, pp. 1764-72, 2001.

[330] R.A. Weiss, "How does HIV cause AIDS?," *Science,* vol. 260, pp. 1273-9, 1993.

[331] H.G. Kräusslich, Ingraham, R.H., Skoog, M.T., Wimmer, E., Pallai, P.V., Carter, C.A., "Activity of purified biosynthetic proteinase of human immunodeficiency virus on natural substrates and synthetic peptides," *PNAS,* vol. 86, pp. 807-11, 1989.

[332] N.E. Kohl, Emini, E.A., Schleif, W.A., Davis, L.J., Heimbach, J.C., Dixon, R.A., Scolnick, E.M., Sigal, I.S., "Active human immunodeficiency virus protease is required for viral infectivity," *PNAS*, vol. 85, pp. 4686-90, 1988.

[333] V. Vapnik, *The Nature of Statistical Learning Theory.* New York: Springer-Verlag, 1995.

[334] Y.D. Cai, Ricardo, P.W., Jen, C.H., Chou, K.C., "Application of SVMs to predict membrane protein types," *Journal of Theoretical Biology*, vol. 226, pp. 373-6, 2004.

[335] K. Park, Kanehisa, M., "Prediction of protein subcellular locations by support vector machines using compositions of amino acids and amino acid pairs," *Bioinformatics*, vol. 19, pp. 1656-63, 2003.

[336] R.J. Carter, Dubchak, I., Holbrook, S.R., "A computational approach to identify genes for functional RNAs in genomic sequences," *Nucleic Acids Res*, vol. 29, pp. 3928-38, 2001.

[337] C.H. Q. Ding, Dubchak, I., "Multi-class protein fold recognition using support vector machines and neural networks," *Bioinformatics*, vol. 17, pp. 349-58, 2001.

[338] K. Lin, Kuang, Y., Joseph, J.S., Kolatkar, P.R., "Conserved codon composition of ribosomal protein coding genes in Escherichia coli, Mycobacterium tuberculosis and Saccharomyces cerevisiae: lessons from supervised machine learning in functional genomics," *NAR*, vol. 30, pp. 2599-2607, 2002.

[339] T. Jaakkola, Diekhans, M., Haussler, D., "A Discriminative Framework for Detecting Remote Protein Homologies," *Journal of Computational Biology*, vol. 7, pp. 95-114, 2000.

[340] R. Karchin, Karplus, K., Haussler, D., "Classifying G-protein coupled receptors with support vector machines," *Bioinformatics*, vol. 18, pp. 147-59, 2002.

[341] Y. Guermeur, Pollastri, G., Elisseeff, A., Zelus, D., Paugam-Moisy, H., Baldi, P., "Combining protein secondary structure prediction models with ensemble methods of optimal complexity," *Neurocomputing*, vol. 56, pp. 305-27, 2004.

[342] L. Liao, Noble, W.S., "Combining pairwise sequence similarity and support vector machines for detecting remote protein evolutionary and structural relationships," *J Comp Biol*, vol. 10, pp. 857-68, 2003.

[343] C.S. Leslie, Eskin, E., Cohen, A., Weston, J., Noble, W.S., "Mismatch string kernels for discriminative protein classification," *Bioinformatics*, vol. 20, pp. 467-76, 2004.

[344] J.J. Ward, Sodhi, J.S., McGuffin, L.J., Buxton, B.F., Jones, D.T., "Prediction and functional analysis of native disorder in proteins from the three kingdoms of life," *J Mol Biol*, vol. 337, pp. 635-45, 2004.

[345] H. Saigo, Vert, J.P., Ueda, N., Akutsu, T., "Protein homology detection using string alignment kernels," *Bioinformatics*, vol. 20, pp. 1682-9, 2004.

[346] A. Zien, Ratsch, G., Mika, S., Scholkopf, B., Lengauer, T. and Muller, K.R., "Engineering support vector machine kernels that recognize translation initiation sites," *Bioinformatics*, vol. 16, pp. 799-807, 2000.

[347]   N. Zavaljevski, Stevens, F.J., Reifman, J., "Support vector machines with selective kernel scaling for protein classification and identification of key amino acid positions," *Bioinformatics,* vol. 18, pp. 689-96, 2002.

[348]   Y. Zhao, Pinilla, C., Valmori, D., Martin, R., Simon, R., "Application of support vector machines for T-cell epitopes prediction," *Bioinformatics,* vol. 19, pp. 1978-84, 2003.

[349]   A. Koike, Takagi, T., "Prediction of protein-protein interaction sites using support vector machines," *Protein Eng Des Sel,* vol. 17, pp. 165-73, 2004.

[350]   M.E. Tipping, "Sparse Bayesian learning and the relevance vector machine," *J Mach Learn Res,* vol. 1, pp. 211-44, 2001.

[351]   D.J. MaCkay, "A practical Bayesian framework for backpropagation networks," *Neural Computation,* vol. 4, pp. 448-72, 1992.

[352]   W. Zhang, Liu, J., Niu, Y.Q., Wang, L., Hu, X., "A Bayesian regression approach to the prediction of MHC-II binding affinity," *Comput Methods Programs Biomed,* vol. 92, pp. 1-7, 2008.

[353]   G.C. Cawley, Talbot, N.L., "Gene selection in cancer classification using sparse logistic regression with Bayesian regularization," *Bioinformatics,* vol. 22, pp. 2348-55, 2006.

[354]   Y. Li, Lee, K.K., Walsh, S., Smith, C., Hadingham, S., Sorefan, K., Cawley, G., Bevan, M.W., "Establishing glucose- and ABA-regulated transcription networks in Arabidopsis by microarray analysis and promoter classification using a Relevance Vector Machine.," *Genome Research,* vol. 16, pp. 414-27, 2006.

[355]   T.A. Down, Hubbard, T.J., "What can we learn from noncoding regions of similarity between genomes?," *BMC Bioinformatics,* vol. 4, p. 131, 2004.

[356]   L.R. Rabiner, "A tutorial on hidden Markov models and selected applications in speech recognition," *Proc. IEEE,* vol. 77, pp. 257-86, 1989.

[357]   D.J. White, *Dynamic Programming.* Edinburgh: Oliver & Boyd, 1969.

[358]   R.E. Bellman, *Dynamic Programming*: Dover Publications, 1962.

[359]   A.J. Viterbi, "Error bounds for convolutional codes and an asymptotically optimum decoding algorithm," *IEEE Transactions on Information Theory,* vol. 13, pp. 260-9, 1967.

[360]   I. Ebersberger, Strauss, S., von Haeseler, A., "HaMStR: profile hidden markov model based search for orthologs in ESTs," *BMC Evol Biol,* vol. 9, p. 157, 2009.

[361]   A. Drawid, Gupta, N., Nagaraj, V.H., Gélinas, C., Sengupta, A.M., "OHMM: a Hidden Markov Model accurately predicting the occupancy of a transcription factor with a self-overlapping binding motif," *BMC Bioinformatics,* vol. 10, p. 208, 2009.

[362]   A.N. Nguyen Ba, Pogoutse, A., Provart, N., Moses, A.M., "NLStradamus: a simple Hidden Markov Model for nuclear localization signal prediction," *BMC Bioinformatics,* vol. 10, p. 202, 2009.

[363]   J.C. Detilleux, "The analysis of disease biomarker data using a mixed hidden Markov model," *Genet Sel Evol,* vol. 40, pp. 491-509, 2008.

[364] X. Deng, Geng, H., Ali, H.H., "A Hidden Markov Model approach to predicting yeast gene function from sequential gene expression data," *Int J Bioinform Res Appl*, vol. 4, pp. 263-73, 2008.

[365] Z.I. Litou, Bagos, P.G., Tsirigos, K.D., Liakopoulos, T.D., Hamodrakas, S.J., "Prediction of cell wall sorting signals in gram-positive bacteria with a hidden markov model: application to complete genomes," *J Bioinform Comput Biol*, vol. 6, pp. 387-401, 2008.

[366] L.E. Baum, Petrie, T., Soules, G., Weiss, N., "A maximization technique occurring in the statistical analysis of probabilistic functions of Markov chains," *Ann. Math. Statist*, vol. 41, pp. 164-71, 1970.

[367] R. Durbin, Eddy, S.R., Krogh, A., Mitchison, G., *Biological Sequence Analysis: Probalistic Models of Proteins and Nucleic Acids*. Cambridge, UK: Cambridge University Press, 1998.

[368] A. Krogh, Brown, M., Mian, I.S., Sjölander, K., Haussler, D., "Hidden Markov models in computational biology. Applications to protein modeling," *J Mol Biol*, vol. 235, pp. 1501-31, 1994.

[369] S.R. Eddy, "A probabilistic model of local sequence alignment that simplifies statistical significance estimation," *PLoS Computational Biology*, vol. 4, p. e1000069, 2008.

[370] H.D. Huang, Lee, T.Y., Tzeng, S.W., Horng, J.T., "KinasePhos: a web tool for identifying protein kinase-specific phosphorylation sites," *NAR*, vol. 33, pp. W226-9, 2005.

[371] Y.H. Wong, Lee, T.Y., Liang, H.K., Huang, C.M., Wang, T.Y., Yang, Y.H., Chu, C.H., Huang, H.D., Ko, M.T., Hwang, J.K., "KinasePhos 2.0: a web server for identifying protein kinase-specific phosphorylation sites based on sequences and coupling patterns," *NAR*, vol. 35, pp. W588-94, 2007.

[372] G. Bejerano, Yona, G., "Variations on probabilistic suffix trees: statistical modeling and prediction of protein families," *Bioinformatics*, vol. 17, pp. 23-43, 2001.

[373] V.C. Epa, "Modeling the paramyxovirus hemagglutinin-neuraminidase protein," *Proteins*, vol. 29, pp. 264-81, 1997.

[374] D. Husmeier, McGuire, G., "Detecting recombination in 4-taxa DNA sequence alignments with Bayesian hidden Markov models and Markov chain Monte Carlo," *Mol Biol Evol*, vol. 20, pp. 315-37, 2003.

[375] P.M. Hooper, Zhang, H., Wishart, D.S., "Prediction of genetic structure in eukaryotic DNA using reference point logistic regression and sequence alignment," *Bioinformatics*, vol. 16, pp. 425-38, 2000.

[376] D.J. Hand, *Discriminant Analysis and Classification*. New York: John Wiley, 1981.

[377] P.A. Devijver, Kittler, J., *Pattern Recognition: A Statistical Approach*. Englewood Cliffs: NJ: Prentice-Hall, 1982.

[378] K. Fuk90aga, *Introduction to Statistical Pattern Recognition*. San Diego: Academic Press, 1982.

[379] A. Webb, *Statistical Pattern Recognition*. Chichester: John Wiley & Sons Ltd, 2002.

[380] R. Tibshirani, "Regression shrinkage and selection via the lasso," *J. Royal. Statist. Soc B*, vol. 58, pp. 267-88, 1996.

[381] T. Hastie, Tibshirani, R., Friedman, J., *The Elements of Statistical Learning*: Springer-Verlag, 2009.

[382] A. Miller, *Subset Selection in Regression*: Chapman & Hall/CRC, 1990.

[383] M.R. Garey, Johnson, D.S., Computers and Intractability: A Guide to the Theory of NP-Completeness: W.H. Freeman, 1979.

[384] T.H. Bradley; E., Iain, J., Robert, T., "Least Angle Regression," *Annals of Statistics*, vol. 32, pp. 407-99, 2004.

[385] I. Sohn, Kim, J., Jung, S.H., Park, C., "Gradient lasso for Cox proportional hazards model," *Bioinformatics*, vol. 25, pp. 1775-81, 2009.

[386] M. Gustafsson, Hörnquist, M., Lundström, J., Björkegren, J., Tegnér, J., "Reverse engineering of gene networks with LASSO and nonlinear basis functions," *Ann N Y Acad Sci*, vol. 1158, pp. 265-75, 2009.

[387] T. Shimamura, Imoto, S., Yamaguchi, R., Miyano, S., "Weighted lasso in graphical Gaussian modeling for large gene network estimation based on microarray data," *Genome Inform*, vol. 19, pp. 142-53, 2007.

[388] W. Shi, Lee, K.E., Wahba, G., "Detecting disease-causing genes by LASSO-Patternsearch algorithm," *BMC Proc*, vol. Suppl 1, p. S60, 2007.

[389] A.N. Tikhonov, Arsenin, V.Y., *Solutions of Ill-Posed Problems*. Washington: Winston, 1977.

[390] A.E. Hoerl, Kennard, R.W., "Ridge regression: Biased estimation for nonorthogonal problems," *Technometrics*, vol. 12, pp. 55-67, 1970.

[391] S. Geman, Bienenstock, E. and Doursat, R., "Neural Networks and the Bias/Variance Dilemma," *Neural Computation*, vol. 4, pp. 1-58, 1992.

[392] P.R. Krishnaiaah, "Multivariate Analysis," New York: Academic Press, 1966.

[393] P. Geladi, Kowlaski B., "Partial least square regression: A tutorial," *Analytica Chemica Acta*, vol. 35, pp. 1-17, 1986.

[394] I. E. Frank, Friedman, J.H., "A statistical view of chemometrics regression tools," *Technometrics*, vol. 35, pp. 109-48, 1993.

[395] I.S. Helland, "Pls regression and statistical models," *Scandivian Journal of Statistics*, vol. 17, pp. 97-114, 1990.

[396] A. Höskuldson, "Pls regression methods," *Journal of Chemometrics*, vol. 2, pp. 211-28, 1988.

[397] A.S. Carpentier, Riva, A., Tisseur, P., Didier, G., Hénaut, A., "The operons, a criterion to compare the reliability of transcriptome analysis tools: ICA is more reliable than ANOVA, PLS and PCA," *Comput Biol Chem*, vol. 28, pp. 3-10, 2004.

[398] A.A. Alaiya, Franzén, B., Hagman, A., Silfverswärd, C., Moberger, B., Linder, S., Auer, G., "Classification of human ovarian tumors using multivariate data analysis of polypeptide expression patterns," *Int J Cancer,* vol. 86, pp. 731-6, 2000.

[399] N.A. Kratochwil, Huber, W., Müller, F., Kansy, M., Gerber, P.R., "Predicting plasma protein binding of drugs: a new approach," *Biochem Pharmacol,* vol. 64, pp. 1355-74, 2002.

[400] Z. Li, Chan, C., "Integrating gene expression and metabolic profiles," *J Biol Chem,* vol. 279, pp. 27124-37, 2004.

[401] U. Rüetschi, Zetterberg, H., Podust, V.N., Gottfries, J., Li, S., Simonsen, A., McGuire, J., Karlsson, M., Rymo, L., Davies, H., Minthon, L., Blennow, K., "Identification of CSF biomarkers for frontotemporal dementia using SELDI-TOF," *Exp Neurol,* vol. 196, pp. 273-81, 2005.

[402] D.V. Nguyen, Rocke, D.M., "On partial least squares dimension reduction for microarray-based classification: a simulation study," *Computational Statistics & Data Analysis,* vol. 46, pp. 407-25, 2004.

[403] I. Guyon, Weston, J., Barnhill, S., Vapnik, V., "Gene selection for cancer classification using support vector machines," *Machine Learning,* vol. 46, pp. 389-422, 2002.

[404] S. Maldonado, Weber, R., "A wrapper method for feature selection using Support Vector Machines," *Information Sciences,* vol. 179, pp. 2208-2217, 2009.

[405] S. Chen, Cowan, C.F.N., Grant, P.M., "Orthogonal least squares learning for radial basis function networks," *IEEE Trans on Neural Networks,* vol. 2, pp. 302-9, 1991.

[406] G.H. Golub, Van Loan, C.F., *Matrix Computations*: Johns Hopkins, 1996.

[407] L.N. Trefethen, Bau, D., *Numerical linear algebra*: Philadelphia: Society for Industrial and Applied Mathematics, 1997.

[408] R.A. Horn, Johnson, C.R., *Matrix Analysis.* Cambridge, UK: Cambridge University Press 1985.

[409] J. Yoo, Patterson, B., Datta, S., "An OLS-based predictor test for a single-index model for predicting transcription rate from histone acetylation level," *Statistics & Probability Letters,* vol. in press.

[410] M.K. Kerr, Martin, M., Churchill, G.A., "Analysis of Variance for Gene Expression Microarray Data," *Journal of Computational Biology,* vol. 7, pp. 819-37, 2000.

[411] C.S. Kim, "Bayesian Orthogonal Least Squares (BOLS) algorithm for reverse engineering of gene regulatory networks," *BMC Bioinformatics,* vol. 8, p. 251, 2007.

[412] T. Suzuki, Sugiyama, M., Kanamori, T., Sese, J., "Mutual information estimation reveals global associations between stimuli and biological processes," *BMC Bioinformatics,* vol. 10, p. S52, 2009.

[413]   S. Szymczak, Nuzzo, A., Fuchsberger, C., Schwarz, D.F., Ziegler, A., Bellazzi, R., Igl, B.W., "Genetic association studies for gene expressions: permutation-based mutual information in a comparison with standard ANOVA and as a novel approach for feature selection," *BMC Proc,* vol. Suppl 1, p. S9, 2007.

[414]   V. Venkatraman, Dalby, A.R., Yang, Z.R., "Evaluation of mutual information and genetic programming for feature selection in QSAR," *J Chem Inf Comput Sci,* vol. 44, pp. 1686-92, 2004.

[415]   A.M. Lesk, *Introduction to Bioinformatics.* Oxford: Oxford University Press, 2008.

[416]   B.L. Theodore, Eugene, L.H., Murphy, B.E., Catherine, M.J., Patrick, W., *Chemistry: The Central Science.* Upper Saddle River, NJ: Pearson/Prentice Hall, 2009.

[417]   H.W. John, Ralph, P.H., Terry, M.W., Scott, P.S., *General Chemistry.* Upper Saddle River, NJ: Pearson/Prentice Hall, 2005.

[418]   W.W. Kenneth, Raymond, D.E., Larry, P.M., *General Chemistry.* Fort Worth, TX: Saunders College Publishing/Harcourt College Publishers, 2000.

[419]   E.A. Hill, "On A System Of Indexing Chemical Literature; Adopted By The Classification Division Of The U. S. Patent Office," *J Am Chem Soc,* vol. 22, pp. 478-94, 1900.

[420]   P.W. Atkins, *The Periodic Kingdom*: HarperCollins Publishers, Inc, 1995.

[421]   R.A. Poorman, Tomasselli, A.G., Heinrikson, R.L., Kezdy, F.J., "A cumulative specificity model for protease from human immunodeficiency virus types 1 and 2, inferred from statistical analysis of an extended substrate data base," *J. Biol. Chem,* vol. 22, pp. 14554-61, 1991.

[422]   T.P. Hopp, Woods, K.R, "A computer program for predicting protein antigenic determinants," *Mol Immunol,* vol. 20, pp. 483-489, 1983.

[423]   J.L. Cornette, Cease, K.B., Margalit, H., Spouge, J.L., Berzofsky, J.A., DeLisi, C., "Hydrophobicity scales and computational techniques for detecting amphipathic structures in proteins," *J Mol Biol,* vol. 195, pp. 687-693, 1987.

[424]   D. Eisenberg, Schwarz, E. Komaromy, M., Wall, R, "Analysis of membrane and surface protein sequences with the hydrophobic moment plot," *J Mol Biol,* vol. 179, pp. 125-133, 1984.

[425]   D. Eisenberg, Weiss, R.M., Terwilliger, T.C, "The hydrophobic moment detects periodicity in protein hydrophobicity," *Proc Natl Acad Sci U S A,* vol. 81, pp. 140-144, 1984.

[426]   D.M. Engelman, Steitz, T.A., Goldman, A, "Identifying nonpolar transbilayer helices in amino acid sequences of membrane proteins," *Annu Rev Biophys Biophys Chem,* vol. 15, pp. 321-323, 1986.

[427]   J. Janin, "Surface and inside volumes in globular proteins," *Nature,* vol. 277, pp. 491-492, 1979.

[428]   G.D.G. Rose, A.R., Lesser, G.J., Lee, R.H., Zehfus, M.H., "Hydrophobicity of amino acid residues in globular proteins," *Science,* vol. 229, pp. 834-838, 1985.

[429] T.E. Creighton, *Proteins: structures and molecular properties.* San Francisco: W.H. Freeman, 1993.

[430] D.W. Urry, "The change in Gibbs free energy for hydrophobic association - Derivation and evaluation by means of inverse temperature transitions," *Chem Phy Lett,* vol. 399, pp. 177-181, 2004.

[431] E. Georges, "The P-glycoprotein (ABCB1) linker domain encodes high-affinity binding sequences to alpha- and beta-tubulins," *Biochemistry,* vol. 46, pp. 7337-7342, 2007.

[432] M. Neuwirth, Flicker, K., Strohmeier, M., Tews, I. and Macheroux, P., "Thermodynamic characterization of the protein-protein interaction in the heteromeric Bacillus subtilis pyridoxalphosphate synthase," *Biochemistry* vol. 46, pp. 5131-5139, 2007.

[433] T. Nomura, Sokabe, M. and Yoshimura, K., "Lipid-Protein Interaction of the MscS Mechanosensitive Channel Examined by Scanning Mutagenesis," *Biophys J,* vol. 91, pp. 2874-2881, 2006.

[434] J. Sohn, Rudolph, J., "Temperature dependence of binding and catalysis for the Cdc25B phosphatase," *Biophys Chem,* vol. 125, pp. 549-555, 2006.

[435] P.H.A. Sneath, "Relations between chemical structure and biological activity in peptides," *J. Theor. Biol,* vol. 12, pp. 157-95, 1966.

[436] R. Grantham, "Amino acid difference formula to help explain protein vvolution," *Science,* vol. 185, pp. 862-64, 1974.

[437] C.D. Livingstone, Barton, G.J., "Protein sequence alignments: a strategy for the hierarchical analysis of residue conservation," *CABIOS,* vol. 9, pp. 745-56, 1993.

[438] G. Mocz, "Fuzzy cluster-analysis of simple physicochemical properties of amino-acids for recognizing secondary structure in proteins," *Protein Sci,* vol. 4, pp. 1178-87, 1995.

[439] L.E. Stanfel, 183, 195-205, "A new approach to clustering the amino acids," *J. Theor. Biol,* vol. 183, pp. 195-205, 1996.

[440] W.R. Taylor, "The classification of amino-acid conservation," *J. Theor. Biol,* vol. 119, pp. 205-18, 1986.

[441] G. Tedeschi, Cappelletti, G., Nonnis, S., Taverna, F., Negri, A., Ronchi, C., Ronchi, S., "Tyrosine nitration is a novel post-translational modification occurring on the neural intermediate filament protein peripherin," *Neurochem Res,* vol. 32, pp. 433-41, 2007.

[442] J.S. Beckman, Ischiropoulos, H., Zhu, L., van der Woerd, M., Smith, C., Chen, J., Harrison, J., Martin, J.C., Tsai, M., "Kinetics of superoxide dismutase- and iron-catalyzed nitration of phenolics by peroxynitrite," *Arch Biochem Biophys,* vol. 298, pp. 438-45, 1992.

[443] H. Ischiropoulos, Zhu, L., Chen, J., Tsai, M., Martin, J.C., Smith, C.D., Beckman, J.S., "Peroxynitrite-mediated tyrosine nitration catalyzed by superoxide dismutase," *Arch Biochem Biophys,* vol. 298, pp. 431-7, 1992.

[444] H. Ischiropoulos, Zhu, L., Beckman, J.S., "Peroxynitrite formation from macrophage-derived nitric oxide," *Arch Biochem Biophys,* vol. 298, pp. 446-51, 1992.

[445] J.S. Beckmann, Ye, Y.Z., Anderson, P.G., Chen, J., Accavitti, M.A., Tarpey, M.M., White, C.R., "Extensive nitration of protein tyrosines in human atherosclerosis detected by immunohistochemistry," *Biol Chem Hoppe Seyler,* vol. 375, pp. 81-8, 1994.

[446] H. Ohshima, Friesen, M., Brouet, I., Bartsch, H., "Nitrotyrosine as a new marker for endogenous nitrosation and nitration of proteins," *Food Chem Toxicol,* vol. 28, pp. 647-52, 1990.

[447] S. PfeifferDagger, Schmidt, K., Mayer, B., "Dityrosine Formation Outcompetes Tyrosine Nitration at Low Steady-state Concentrations of Peroxynitrite - IMPLICATIONS FOR TYROSINE MODIFICATION BY NITRIC OXIDE/SUPEROXIDE IN VIVO," *J Biol Chem,* vol. 275, pp. 6346-6352, 2000.

[448] D.D. Thomas, Espey, M.G., Vitekm, M.P., Miranda, K.M., Wink, D.A., "Protein nitration is mediated by heme and free metals through Fenton-type chemistry: An alternative to the NO/OFormula reaction," *PNAS,* vol. 99, pp. 12691-12696, 2002.

[449] J.P. Eiserich, Hristova, M., Cross, C.E., Jones, A.D., Freeman, B.A., Halliwell, B., van der Vliet, A., "Formation of nitric oxide-derived inflammatory oxidants by myeloperoxidase in neutrophils," *Nature,* vol. 391, pp. 393-7, 1998.

[450] K. Bian, Gao, Z., Weisbrodt, N., Murad, F., "The nature of heme/iron-induced protein tyrosine nitration," *PNAS,* vol. 100, pp. 5712-5717, 2003.

[451] M.L. Brennan, Wu, W., Fu, X., Shen, Z., Song, W., Frost, H., Vadseth, C., Narine, L., Lenkiewicz, E., Borchers, M.T., Lusis, A.J., Lee, J.J., Lee, N.A., Abu-SoudDagger, H.M., Ischiropoulos, H., Hazen, S.L., "A Tale of Two Controversies, defining both the role of peroxidases in nitrotyrosine formation in vivo using eosinophil peroxidase and myeloperoxidase-deficient mice, and the nature of peroxidase-generated reactive nitrogen species," *J. Biol. Chem,* vol. 277, pp. 17415-17427, 2002.

[452] T. Elsasser, Kahl, S., Sartin, J., Li, C., "Protein tyrosine nitration: a membrane-organized mechanism for altered signal transduction during proinflammatory stress," *American Society of Animal Science,* vol. 82, p. 445, 2004.

[453] G. Cappelletti, Maggioni, M.G., Ronchi, C., Maci, R., Tedeschi, G., "Protein tyrosine nitration is associated with cold- and drug-resistant microtubules in neuronal-like PC12 cells," *Neurosci Lett,* vol. 401, pp. 159-64, 2006.

[454] G. Cappelletti, Maggioni, M.G., Tedeschi, G., Maci, R., "Protein tyrosine nitration is triggered by nerve growth factor during neuronal differentiation of PC12 cells," *Exp Cell Res,* vol. 288, pp. 9-20, 2003.

[455] A. Iwagaki, Choe, N., Li, Y., Hemenway, D.R., Kagan, E., "Asbestos inhalation induces tyrosine nitration associated with extracellular signal-regulated kinase 1/2 activation in the rat lung," *Am J Respir Cell Mol Biol,* vol. 28, pp. 51-60, 2003.

[456] K. Camphausen, Tofilon, P.J., "Inhibition of histone deacetylation: a strategy for tumor radiosensitization," *J Clin Oncol*, vol. 25, pp. 4051-4056, 2007.

[457] L.M. Iakoucheva, Radivojac, P., Brown, C.J., O'Connor, T.R., Sikes, J.G., Obradovic, Z., Dunker, A.K., " Intrinsic disorder and protein phosphorylation," *Nucleic Acids Research*, vol. 32, pp. 1037-1049, 2004.

[458] R. Radi, "Nitric oxide, oxidants, and protein tyrosine nitration," *PNAS*, vol. 101, pp. 4003-4008, 2004.

[459] J.M. Souza, Peluffo, G., Radi, R., "Protein tyrosine nitration-Functional alteration or just a biomarker?," *Free Radic Biol Med*, vol. in press, 2009.

[460] J.M. Van Zyl, Van der Walt, B.J., "Apparent hydroxyl radical generation without transition metal catalysis and tyrosine nitration during oxidation of the anti-tubercular drug, isonicotinic acid hydrazide," *Biochem Pharmacol*, vol. 48, pp. 2033-42, 1994.

[461] V. Villard, Espallergues, J., Keller, E., Alkam, T., Nitta, A., Yamada, K., Nabeshima, T., Vamvakides, A., Maurice, T., "Antiamnesic and neuroprotective effects of the aminotetrahydrofuran derivative ANAVEX1-41 against amyloid beta(25-35)-induced toxicity in mice," *Neuropsychopharmacology*, vol. 34, pp. 1552-66, 2009.

[462] S. Doublier, Riganti, C., Voena, C., Costamagna, C., Aldieri, E., Pescarmona, G., Ghigo, D., Bosia, A., "RhoA silencing reverts the resistance to doxorubicin in human colon cancer cells "*Mol Cancer Res*, vol. 6, pp. 1607-20, 2008.

[463] D.E. Rumelhart, McClelland, J.L., *Parallel Distributed Processing - Vol. 1: Foundations*: The MIT Press, 1987.

[464] M.V. Brock, Gou, M., Akiyama,Y., Muller, A., Wu, T., Montgomery, E., Deasel, M., Germonpré, P., Rubinson, L., Heitmiller, R.F., Yang, S.C., Forastiere, A.A., Baylin, S.B., Herman, J.G., "Prognostic importance of promoter hypermethylation of multiple genes in esophageal adenocarcinoma," *Clinical Cancer Research*, vol. 9, pp. 2912-9, 2003.

[465] D.L. Mandelker, Yamashita, K., Tokumaru, Y., Mimori, K., Howard, D.L., Tanaka, Y., Carvalho, A.L., Jiang, W., Park, H., Kim, M., Osada,M., Mori, M., Sidransky, D., "PGP9.5 Promoter Methylation Is an Independent Prognostic Factor for Esophageal Squamous Cell Carcinoma " *Cancer Res*, vol. 65, pp. 4963-8, 2005.

[466] M.Q. Hoque, Rosenbaum, E., Westra, W.H., Xing, M., Ladenson, P., Zeiger, M.A., Sidransky, D., Umbricht, C.B., "Quantitative Assessment of Promoter Methylation Profiles in Thyroid Neoplasms," *J. Clin. Endocrinol. Metab*, vol. 90, pp. 4011-8, 2005.

[467] R. Martinez, Setien, F., Voelter, C., Casado, S., Quesada, M.P., Schackert, G., Esteller, M., "CpG island promoter hypermethylation of the pro-apoptotic gene caspase-8 is a common hallmark of relapsed glioblastoma multiforme," *Carcinogenesis*, vol. 28, pp. 1264-8, 2007.

[468]	D.K. Hawley, McClure, W.R., "Compilation and analysis of Escherichia coli promoter DNA sequences," *NAR,* vol. 11, pp. 2237-55, 1983.

[469]	M. Rosenberg, Court, D., "Regulatory sequences involved in the promotion and termination of RNA transcription," *Annu Rev Genet,* vol. 13, pp. 319-53, 1979.

[470]	J.D. Helmann, deHaseth, P.L., "Protein-nucleic acid interactions during open complex formation investigated by systematic alteration of the protein and DNA binding partners," *Biochemistry,* vol. 38, pp. 5959-67, 1999.

[471]	A. Kanhere, Bansal, M., "A novel method for prokaryotic promoter prediction based on DNA stability," *BMC Bioinformatics,* vol. 6, p. 1, 2005.

[472]	V. Cotik, Zaliz, R.R., Zwir, I., "A hybrid promoter analysis methodology for prokaryotic genomes," *Fuzzy Sets and Systems,* vol. 152, pp. 83-102, 2004.

[473]	H.B. Wang, C.J., "Promoter prediction and annotation of microbial genomes based on DNA sequence and structural responses to superhelical stress," *BMC Bioinformatics,* vol. 7, p. 248, 2006.

[474]	I. Gershenzon. N.I., I.P., "Synergy of human Pol II core promoter elements revealed by statistical sequence analysis," *Bioinformatics,* vol. 21, pp. 1295-300, 2005.

[475]	V.B. Bajic, Brent, M.R., Brown, R.H., Frankish, A., Harrow, J., Ohler, U., Solovyev, V.V., Tan, S.L., "Performance assessment of promoter predictions on ENCODE regions in the EGASP experiment," *Genome Biol,* vol. 7, pp. S3.1-S3.13, 2006.

[476]	S. Sonnenburg, Zien, A., Rätsch, G., "ARTS: accurate recognition of transcription starts in human "*Bioinformatics,* vol. 22, pp. e472-e480, 2006.

[477]	J. Wang, Ungar, L.H., Tseng, H., Hannenhalli, S., "MetaProm: a neural network based meta-predictor for alternative human promoter prediction," *BMC Bioinformatics,* vol. 8, p. 374, 2007.

[478]	X. Xie, Wu, S., Lam, K., Yan, H., "PromoterExplorer: an effective promoter identification method based on the AdaBoost algorithm," *Bioinformatics,* vol. 22, pp. 2722-8, 2006.

[479]	T. Abeel, Saeys, Y., Rouzé, P., de Peer, Y.V., "ProSOM: core promoter prediction based on unsupervised clustering of DNA physical profiles," *Bioinformatics,* vol. 24, pp. i24-i31, 2008.

[480]	P. Römer, Hahn, S., Jordan, T., Strauss, T., Bonas, U., Lahaye, T., "Plant pathogen recognition mediated by promoter activation of the pepper Bs3 resistance gene," *Science,* vol. 318, pp. 645-8, 2007.

[481]	S.D. Soby, Daniels, M.J., "Catabolite-repressor-like protein regulates the expression of a gene under the control of the Escherichia coli lac promoter in the plant pathogen Xanthomonas campestris pv. campestris," *Appl Microbiol Biotechnol,* vol. 46, pp. 559-61, 1996.

[482]	P. Siriputthaiwan, Jauneau, A., Herbert, C., Garcin, D., Dumas, B., "Functional analysis of CLPT1, a Rab/GTPase required for protein secretion and pathogenesis

in the plant fungal pathogen Colletotrichum lindemuthianum," *J Cell Sci*, vol. 118, pp. 323-9, 2005.

[483] S. Robatzek, Somssich, I.E., "Targets of AtWRKY6 regulation during plant senescence and pathogen defense," *Genes Dev*, vol. 16, pp. 1139-49, 2002.

[484] P.J. Rushton, Reinstädler, A., Lipka, V., Lippok, B., Somssich, I.E., "Synthetic plant promoters containing defined regulatory elements provide novel insights into pathogen- and wound-induced signaling," *Plant Cell*, vol. 14, pp. 749-62, 2002.

[485] I.A. Shahmuradov, Gammerman, A.J., Hancock, J.M., Bramley, P.M., Solovyev, V.V., "PlantProm: a database of plant promoter sequences," *NAR*, vol. 31, pp. 114-7, 2003.

[486] I.A. Shahmuradov, Solovyev, V.V., Gammerman, A.J., "Plant promoter prediction with confidence estimation," *NAR*, vol. 33, pp. 1069-76, 2005.

[487] F. Anwar, Baker, S.M., Jabid, T., Mehedi Hasan, M., Shoyaib, M., Khan, H., Walshe, R., "Pol II promoter prediction using characteristic 4-mer motifs: a machine learning approach," *BMC Bioinformatics*, vol. 9, p. 414, 2008.

[488] J.M. Ostell, Kans, J.A., "The NCBI data model," *Methods Biochem Anal*, vol. 39, pp. 121-44, 1998.

[489] F. Anwar, Baker, S., Jabid, T., Hasan, M.M., Shoyaib, M., Khan, H., Walshe, R., "Pol II promoter prediction using characteristic 4-mer motifs: a machine learning approach," *BMC Bioinformatics*, vol. 9, pp. 1-8, 2008.

[490] M. Kanehisa, *Post-genome Informatics*. Oxford: Oxford University Press, 2000.

[491] J. Zhou, Tang, X., Martin, G.B., "The Pto kinase conferring resistance to tomato bacterial speck disease interacts with proteins that bind a cis-element of pathogenesis-related genes," *The EMBO Journal*, vol. 16, pp. 3207-18, 1997.

[492] G.W. Hatfield, Hung, S.P., Baldi, P., "Differential analysis of DNA microarray gene expression data," *Mol Microbiol*, vol. 47, pp. 871-7, 2003.

[493] N. Pavelka, Pelizzola, M., Vizzardelli, C., Capozzoli, M., Splendiani, A., Granucci, F., Ricciardi-Castagnoli, P., "A power law global error model for the identification of differentially expressed genes in microarray data," *BMC Bioinformatics*, vol. 5, p. 203, 2004.

[494] N.M. Luscombe, Qian, J., Zhang, Z., Johnson, T., Gerstein, M., "The dominance of the population by a selected few: power-law behaviour applies to a wide variety of genomic properties," *Genome Biol*, vol. 3, p. 8, 2002.

[495] C. Christensen, Gupta, A., Maranas, C.D., Albert, R., "Large-scale inference and graph-theoretical analysis of gene-regulatory networks in B. Subtilis," *Physica A: Statistical and Theoretical Physics*, vol. 373, pp. 796-810.

[496] T. Akutsu, Kuhara, S., Maruyama, O., Miyano, S., "Identification of genetic networks by strategic gene disruptions and gene overexpressions under a boolean model," *Theoretical Computer Science*, vol. 298, pp. 235-51.

[497] T. Yang, Zhang, L., Zhang, T., Zhang, H., Xu, S., An, L., "Transcriptional regulation network of cold-responsive genes in higher plants," *Plant Science*, vol. 169, pp. 987-95, 2005.

[498] D. Croes, Couche, F., Wodak, S.J., van Helden, J., "Inferring Meaningful Pathways in Weighted Metabolic Networks," *Journal of Molecular Biology*, vol. 356, pp. 222-36, 2006.

[499] O. Resendis-Antonio, Freyre-González, J.A., Menchaca-Méndez, R., Gutiérrez-Ríos, R.M., Martínez-Antonio, A., Ávila-Sánchez, C., Collado-Vides, J., "Modular analysis of the transcriptional regulatory network of E. coli," *Trends in Genetics*, vol. 21, pp. 16-20, 2005.

[500] L. de Campos, Gomes, T., Von Zuben, F.J., Moscato, P., "A proposal for direct-ordering gene expression data by self-organising maps," *Applied Soft Computing*, vol. 5, pp. 11-21, 2004.

[501] J. Gómez-Gardeñes, Moreno, Y., Floría, L.M., "On the robustness of complex heterogeneous gene expression networks," *Biophysical Chemistry*, vol. 115, pp. 225-8, 2005.

[502] M. Aldana, Balleza, E., Kauffman, S., Resendiz, O., "Robustness and evolvability in genetic regulatory networks," *Journal of Theoretical Biology*, vol. 245, pp. 433-48, 2007.

[503] J. Aracena, Goles, E., Moreira, A., Salinas, L., "On the robustness of update schedules in Boolean networks," *Biosystems*, vol. 97, pp. 1-8, 2009.

[504] R.E. Neapolitan, *Learning Bayesian Networks*. Upper Saddle River, NJ: Prentice Hall, 2004.

[505] F. Jensen, *Bayesian Networks and Decision Graphs*: Springer-Verlag, 2001.

[506] C.J. Robert, "Concepts of Independence for Proportions with a Generalization of the Dirichlet Distribution," *Journal of the American Statistical Association*, vol. 64, pp. 194-206, 1969.

[507] E. Steele, Tucker, A., 't Hoen, P.A., Schuemie, M.J., "Literature-based priors for gene regulatory networks," *Bioinformatics*, vol. 25, pp. 1768-74, 2009.

[508] P. Li, Zhang, C., Perkins, E.J., Gong, P., Deng, Y., "Comparison of probabilistic Boolean network and dynamic Bayesian network approaches for inferring gene regulatory networks," *BMC Bioinformatics*, vol. Suppl 7, p. S13, 2007.

[509] X. W. Chen, Anantha, G., Wang, X., "An effective structure learning method for constructing gene networks," *Bioinformatics*, vol. 22, pp. 1367-74, 2006.

[510] Y. Ko, Zhai, C., Rodriguez-Zas, S., "Inference of gene pathways using mixture Bayesian networks," *BMC Syst Biol*, vol. 3, p. 54, 2009.

[511] S.L. Rodriguez-Zas, Ko, Y., Adams, H.A., Southey, B.R., "Advancing the understanding of the embryo transcriptome co-regulation using meta-, functional, and gene network analysis tools," *Reproduction*, vol. 135, pp. 213-24, 2008.

[512] J.M. Peña, Björkegren, J., Tegnér, J., "Growing Bayesian network models of gene networks from seed genes," *Bioinformatics*, vol. Suppl 2, pp. ii224-9, 2005.

[513] S. Kim, Imoto, S., Miyano, S., "Dynamic Bayesian network and nonparametric regression for nonlinear modeling of gene networks from time series gene expression data," *Biosystems,* vol. 75, pp. 57-65, 2004.

[514] J. Cheng, Greiner, R., Kelly, J., Bell, D.A., Liu, W., "Learning Bayesian networks from data: an information-theory based approach," *The Artificial Intelligence Journal,* vol. 137, pp. 43-90, 2002.

[515] M.L. Wong, Lee, S.Y., Leung, K.S., "Data mining of Bayesian networks using cooperative coevolution," *Decision Support Systems,* vol. 38, p. 3, 2004.

[516] N. Friedman, Linial, M., Nachman, I., Pe'er, D., "Using Bayesian network to analyze expression data," *Journal of Computational Biology,* vol. 7, pp. 601-20, 2000.

[517] Z. Huang, Li, J., Su, H., Watts, G.S., Chen, H., "Large-scale regulatory network analysis from microarray data: modified Bayesian network learning and association rule mining," *Decision Support Systems,* vol. 43, pp. 1207-25, 2007.

[518] E. Steele, Tucker, A., "Consensus and Meta-analysis regulatory networks for combining multiple microarray gene expression datasets," *Journal of Biomedical Informatics,* vol. 41, pp. 914-26, 2008.

[519] S.P. Li, Tseng, J.J., Wang, S.C., "Reconstructing gene regulatory networks from time-series microarray data," *Physica A: Statistical Mechanics and its Applications,* vol. 350, pp. 63-9, 2005.

[520] A. Narayanan, Wu, X., Yang, Z.R., "Mining viral protease data to extract cleavage knowledge," *Bioinformatics* vol. 18, pp. S5-13, 2002.

[521] M.A. Savageau, "Biochemical systems analysis. I. Some mathematical properties of the rate law for the component enzymatic reactions," *J. Theor. Biol,* vol. 25, pp. 365-9, 1969.

[522] M.A. Savageau, "Biochemical systems analysis. II. The steady-state solutions for an n-pool system using a power-law approximation," *J Theor Biol,* vol. 25, pp. 370-9, 1969.

[523] M.A. Savageau, "Biochemical systems analysis. III. Dynamic solutions using a power-law approximation," *J Theor Biol,* vol. 26, pp. 215-26, 1970.

[524] L. Michaelis, Menten, M.L., "Die kinetik der invertinwirkung," *Biochem, Zeitschrift,* vol. 49, pp. 333-69, 1913.

[525] E.O. Voit, Computational Analysis of Biochemical Systems, A Practical Guide for Biochemists and Molecular Biologist. Cambridge: Cambridge University Press, 2000.

[526] J.S. Almeida, Voit, E.O., "Neural-network-based parameter estimation in S-system models of biological networks," *Genome Informatics,* vol. 14, pp. 114-23, 2003.

[527] E.O. Voit, Almeida, J., "Decoupling dynamical systems for pathway identification from metabolic profiles," *Bioinformatics,* vol. 20, pp. 1670-81, 2004.

[528] P. Mendes, Kell, D.B., "On the analysis of the inverse problem of metabolic pathways using artificial neural networks," *Biosystems,* vol. 38, pp. 15-28, 1996.

[529] W.H. Press, Flannery, B.P., Teukolsky, S.A., Vetterling, W.T., *Numerical Recipes in C: The Art of Scientific Computing.* Cambridge: Cambridge University Press, 1992.

[530] O.R. Gonzalez, Kuper, C., Jung, K., Naval, P.C., Mendoza, E., "Parameter estimation using simulated annealing for S-system models of biochemical networks," *Bioinformatics,* vol. 23, pp. 480-6, 2007.

[531] S. Kirkpatrick, Gelatt, C.D., Jr Vecchi, M.P., "Optimization by simulated annealing," *Science,* vol. 220, p. 4598, 1983.

[532] J.H. Holland, *Adaptation in Natural and Artificial Systems.* Ann Arbor: University of Michigan Press, 1975.

[533] D.E. Goldberg, Genetic algorithms in search, optimization and machine learning: Addison Wesley, 1989.

[534] S. Kikuchi, Tominaga, D., Arita, M., Takahashi, K., Tomita, M., "Dynamic modelling of genetic networks using genetic algorithm and S-system," *Bioinformatics,* vol. 19, pp. 643-50, 2003.

[535] K. Edwards, Edgar, T.F., Manousiouthankis, V.I., "Kinetic model reduction using genetic algorithms," *Comput Chem Engng,* vol. 22, pp. 239-46, 1998.

[536] S. Kimura, Ide, K., Kashihara, A., Kano, M., Hatakeyama, M., Masui, R., Nakagawa, N., Yokoyama, S., Kuramitsu, S., Konagaya, A., "Inference of S-system models of genetic networks using a cooperative coevolutionary algorithm," *Bioinformatics,* vol. 21, pp. 1154-63, 2005.

[537] P. Liu, Wang, F., "Inference of biochemical network models in S-systems using multiobjective optimisation approach," *Bioinformatics,* vol. 24, pp. 1085-92, 2008.

[538] N. Blow, "Metabolomics: Biochemistry's new look," *Nature,* vol. 455, pp. 697-700, 2008.

[539] M. Herrero, Ibáñez, E., Cifuentes, A., Bernal, J., "Multidimensional chromatography in food analysis," *J Chromatogr A,* vol. in press, 2009.

[540] S. Rochat, Egger, J., Chaintreau, A., "Strategy for the identification of key odorants: application to shrimp aroma," *J Chromatogr A,* vol. 1216, pp. 6424-32, 2009.

[541] Z. Luo, Heffner, C., Solouki, T., "Multidimensional GC-fourier transform ion cyclotron resonance MS analyses: utilizing gas-phase basicities to characterize multicomponent gasoline samples," *J Chromatogr Sci,* vol. 47, pp. 75-82, 2009.

[542] L. Cai, Koziel, J.A., Dharmadhikari, M., van Leeuwen, H.J., "Rapid determination of trans-resveratrol in red wine by solid-phase microextraction with on-fiber derivatization and multidimensional gas chromatography-mass spectrometry," *J Chromatogr A,* vol. 1216, pp. 281-7, 2009.

[543] H. Hühnerfuss, Shah, M.R., "Enantioselective chromatography-a powerful tool for the discrimination of biotic and abiotic transformation processes of chiral environmental pollutants," *J Chromatogr A*, vol. 1216, pp. 481-502, 2009.

[544] P.Q. Tranchida, Costa, R., Donato, P., Sciarrone, D., Ragonese, C., Dugo, P., Dugo, G., Mondello, L., "Acquisition of deeper knowledge on the human plasma fatty acid profile exploiting comprehensive 2-D GC," *J Sep Sci*, vol. 31, pp. 3347-51, 2008.

[545] D.S. Wishart, Lewis, M.J., Morrissey, J.A., Flegel, M.D., Jeroncic, K., Xiong, Y., Cheng, D., Eisner, R., Gautam, B., Tzur, D., Sawhney, S., Bamforth, F., Greiner, R., Li, L:, "The human cerebrospinal fluid metabolome," *J Chromatogr B Analyt Technol Biomed Life Sci*, vol. 87, pp. 164-73, 2008.

[546] T.M. Ebbels, Buxton, B.F., Jones, D.T., "springScape: visualisation of microarray and contextual bioinformatic data using spring embedding and an 'information landscape'," *Bioinformatics*, vol. 22, pp. e99-107, 2006.

[547] T. Lombardot, Kottmann, R., Giuliani, G., de Bono, A., Addor, N., Glackner, F.O., "MetaLook: a 3D visualisation software for marine ecological genomics," *BMC Bioinformatics*, vol. 8, p. 406, 2007.

[548] B.J. Holden, Pinney, J.W., Lovell, S.C., Amoutzias, G.D., Robertson, D.L., "An exploration of alternative visualisations of the basic helix-loop-helix protein interaction network," *BMC Bioinformatics*, vol. 8, p. 289, 2007.

[549] T.C. Freeman, Goldovsky, L., Brosch, M., van Dongen, S., Maziare, P., Grocock, R.J., Freilich, S., Thornton, J., Enright, A.J., "Construction, visualisation, and clustering of transcription networks from microarray expression data," *PLoS Computational Biology*, vol. 10, pp. 2032-42, 2007.

[550] K. O'Neill, Garcia, A., Schwegmann, A., Jimenez, R.C., Jacobson, D., Hermjakob, H., "OntoDas - a tool for facilitating the construction of complex queries to the Gene Ontology," *BMC Bioinformatics*, vol. 9, p. 437, 2008.

[551] T.D. Smith, Cotton, R.G., "VariVis: a visualisation toolkit for variation databases," *BMC Bioinformatics*, vol. 9, p. 206, 2008.

[552] M. Steinfatha, Grotha, D., Lisecb, J., Selbig, J., "Metabolite profile analysis: from raw data to regression and classification," *Physiologia Plantarum*, vol. 132, pp. 150-61, 2008.

[553] W. Weckwerth, "Integration of metabolomics and protepmics in molecular plant physiology – coping with the complexity by data-dimensionality reduction," *Physiology Plantarum*, vol. 132, pp. 176-89, 2008.

[554] E.J. Cooke, Savage, R.S., Wild, D.L., "Computational approaches to the integration of gene expression, ChIP-chip and sequence data in the inference of gene regulatory networks," *Seminars in Cell & Developmental Biology*, vol. in press, 2009.

[555] M. Hecker, Lambeck, S., Toepfer, S., van Someren, E., Guthke, R., "Gene regulatory network inference: Data integration in dynamic models—A review," *Biosystems*, vol. 96, pp. 86-103, 2009.

[556] D. Cavalieri, De Filippo, C., "Bioinformatic methods for integrating whole-genome expression results into cellular networks," *Drug Discovery Today,* vol. 10, pp. 727-34, 2005.

[557] S. Jung, Lee, K.H., Lee, D., "H-CORE: Enabling genome-scale Bayesian analysis of biological systems without prior knowledge," *Biosystems,* vol. 90, pp. 197-210, 2007.

[558] D. Allen, Darwiche, A., "RC_Link: Genetic linkage analysis using Bayesian networks," *International Journal of Approximate Reasoning,* vol. 48, pp. 499-525, 2008.

[559] W. Liao, Ji, Q., "Learning Bayesian network parameters under incomplete data with domain knowledge," *Pattern Recognition,* vol. 42, pp. 3046-56, 2009.

[560] A.A. Alizadeh, Eisen, M.B., Davis, R.E., Ma, C., Lossos, I.S., et al., "Distinct types of diffuse large B-cell lymphoma identified by gene expression profiling," *Nature,* vol. 403, pp. 503-11, 2000.

[561] S.D. Golub TR, Tamayo P, Huard C, Gaasenbeek M, et al., "Molecular classification of cancer: class discovery and class prediction by gene expression monitoring," *Science,* vol. 286, pp. 531-7, 1999.

[562] S. Ramaswamy, Ross, K.N., Lander, E.S., Golub, T.R., "A molecular signature of metastasis in primary solid tumors," *Nat Genet,* vol. 33, pp. 49-54, 2003.

[563] D.H. van 't Veer L.J., van de Vijver M.J., He Y.D., Hart A.A., et al., "Gene expression profiling predicts clinical outcome of breast cancer," *Nature,* vol. 415, pp. 530-6, 2002.

[564] K. Kadota, Nakai, Y., Shimizu, K. (2009). Algorithms Mol Biol. 22; 4:7, "Ranking differentially expressed genes from Affymetrix gene expression data: methods with reproducibility, sensitivity, and specificity," *Algorithms Mol Biol,* vol. 4, p. 7, 2009.

[565] Y. Saeys, Inza, I. and Larrañaga, P., "A review of feature selection techniques in bioinformatics," *Bioinformatics,* vol. 23, pp. 2507-17, 2007.

[566] V.G. Tusher, Tibshirani, R., Chu, G., "Significance analysis of microarrays applied to the ionizing radiation response," *PNAS,* vol. 98, pp. 5116-21, 2001.

[567] L.A.D. Baldi P., "A Bayesian framework for the analysis of microarray expression data: regularized t-test and statistical inferences of gene changes," *Bioinformatics,* vol. 17, pp. 509-19, 2001.

[568] J. Zhang, Finney, R.P., Rowe, W., Edmonson, M., Yang, S.H., et al., "Systematic analysis of genetic alterations in tumors using Cancer Genome WorkBench (CGWB)," *Genome Research,* vol. 17, pp. 1111-7, 2007.

[569] D.J. Slamon, Clark, G.M., Wong, S.G., Levin, W.J., Ullrich, A., McGuire, W.L., "Human breast cancer: correlation of relapse and survival with amplification of the HER-2/neu oncogene," *Science,* vol. 235, pp. 177-182, 1987.

[570] S.A. e. a. Tomlins, "Chinnaiyan, Recurrent fusion of TMPRSS2 and ETS transcription factor genes in prostate cancer," *Science,* vol. 310, pp. 644-8, 2005.

[571] R. Tibshirani, Hastie, T., "Outlier sums for differential gene expression analysis," *Biostatistics,* vol. 8, pp. 2-8, 2007.

[572] H. Lian, "MOST: detecting cancer differential gene expression," *Biostatistics,* vol. 9, pp. 411-8, 2008.

[573] W.N. van Wieringen, van de Wiel, M.A., van der vaart, A.W., "A test for partial differential expression," *Journal of the American Statistical Association,* vol. 103, pp. 1039-49, 2008.

Index